Ergebnisse der Mathematik und ihrer Grenzgebiete

Band 54

Herausgegeben von
P. R. Halmos · P. J. Hilton · R. Remmert · B. Szőkefalvi-Nagy

Unter Mitwirkung von
L. V. Ahlfors · R. Baer · F. L. Bauer · R. Courant · A. Dold
J. L. Doob · S. Eilenberg · M. Kneser · G. H. Müller · M. M. Postnikov
B. Segre · E. Sperner

Geschäftsführender Herausgeber: P. J. Hilton

Herbert Busemann

Recent Synthetic Differential Geometry

Springer-Verlag New York · Heidelberg · Berlin 1970

Herbert Busemann
Professor of Mathematics
University of Southern California
Los Angeles, California 90007

This work was partially supported by National Science Foundation Grant
No. GP-7472

ISBN-13: 978-3-642-88059-9 e-ISBN-13: 978-3-642-88057-5
DOI: 10.1007/978-3-642-88057-5

Title No. 4598
Softcover reprint of the hardcover 1st edition 1970

Preface

A synthetic approach to intrinsic differential geometry in the large and its connections with the foundations of geometry was presented in "The Geometry of Geodesics" (1955, quoted as *G*). It is the purpose of the present report to bring this theory up to date. Many of the later investigations were stimulated by problems posed in *G*, others concern new topics.

Naturally references to *G* are frequent. However, large parts, in particular Chapters I and III as well as several individual sections, use only the basic definitions. These are repeated here, sometimes in a slightly different form, so as to apply to more general situations. In many cases a quoted result is quite familiar in Riemannian Geometry and consulting *G* will not be found necessary.

There are two exceptions: The theory of parallels is used in Sections 13, 15 and 17 without reformulating all definitions and properties (of co-rays and limit spheres). Secondly, many items from the literature in *G* (pp. 409–412) are used here and it seemed superfluous to include them in the present list of references (pp. 106–110). The quotations are distinguished by [ ] and ⟨ ⟩, so that, for example, Freudenthal [1] and ⟨1⟩ are found, respectively, in *G* and here.

Among the research topics suggested in *G* is the extension of the theory to nonsymmetric distances. Completeness can then be given a strong and a weak form. Zaustinsky ⟨1⟩ showed that with strong completeness the methods for the symmetric case carry over with surprisingly few changes, although the formulations of theorems and definitions often become more involved.

It seems that under the weaker completeness hypothesis, for which there are very interesting examples, major revisions become necessary, but systematic investigations do not yet exist[1].

Therefore we restrict ourselves, except for Sections 1, 2, 8, and 12 to the symmetric case, indicating here and there the modifications required for a nonsymmetric distance.

It may be well to emphasize here again that the principal merit of our theory is not the absence of differentiability, although this does lead to more geometric arguments and establishes the very appealing link with the foundations of geometry. We would claim, indeed, that our methods

[1] Such research is in progress now, see the note at the end of the text, p. 104.

have yielded a much larger body of theorems in the large on Finsler spaces and of a much greater variety than any other. It is true that considering a Finsler space essentially as a Riemannian metric defined on the unit tangent bundle has led to the extension of important Riemannian theorems. These excursions into limited areas of Finsler space theory with typically Riemannian methods yielding strict generalization of results in, but no additional information on, Riemannian geometry, do not establish their value in novel, in particular typically non-Riemannian situations, which, in our opinion, offer the real challenge.

In the last part VI, "Observations on Method and Content," we have enlarged on these views, but do not expect that everyone will concur.

Herbert Busemann

Contents

I. Completeness, Finite Dimensionality, Differentiability

The first chapter treats questions concerning the foundations of the theory which were either not asked in G or left open.

The completeness concepts entering differential geometry were analysed in the well known paper of Hopf and Rinow ⟨1⟩. Soon afterwards Cohn-Vossen ⟨1⟩ recognized that their main result is independent of differentiability assumptions. The, or a, *Hopf-Rinow Theorem* is found in almost all recent books on advanced differential geometry, but in a form which is open to objections: in order to obtain a comprehensive theorem, hypotheses are introduced which are irrelevant to the principal conclusion (also under smoothness hypotheses) or which make essentially different conditions coincide. The first two sections clarify the issues.

We then modify the *theory of r-spaces by Kosiński* ⟨1⟩ so that it applies to G-spaces (the axioms for these spaces are listed in G, p. 37 and p. 14 here). It yields that *finite-dimensional G-spaces have important topological properties such as domain invariance and the noncontractibility of small spheres.* The latter is applied to problems concerning conjugate points.

Finally we ask how one can *recognize from the distance of a G-space whether it is a Finsler space*, i.e., obtainable from a line element on a differentiable manifold satisfying the standard conditions (see G, Section 15). The answer is surprisingly simple although the proofs are long, and is interesting not only in the framework of our theory, but also from the classical point of view. A distance pq on a manifold may fail to be smooth as a function of the local coordinates of p and q. (For example, pq may stem from a Finsler space, but be written in terms of nonadmissible coordinates.) How can one decide whether a more favorable choice of coordinates exists?

1. The Theorem of Hopf and Rinow

Before entering the discussion of completeness we recapitulate some basic concepts and notations.

The *axioms* for a metric space R with a *not necessarily symmetric distance* are: *a real valued function* xy *is defined on* $R \times R$ *with the properties*

$$xx=0, \quad xy>0 \quad \textit{for } x \neq y, \quad xy+yz \geqq xz$$

and

$$x x_\nu \to 0 \quad \textit{if and only if} \quad x_\nu x \to 0.$$

(For this and the following see Busemann [3].) Since

$$\delta(x, y) = \max(x y, y x)$$

satisifes the conditions for an ordinary metric space and $\delta(x, x_\nu) \to 0$ is equivalent to $x x_\nu \to 0$ or $x_\nu x \to 0$ we use the topology induced by $\delta(x, y)$.

For *spherical neighborhoods and spheres of radius* $\rho > 0$ we use the notations

$$S^+(p, \rho) = \{x \mid p x < \rho\}, \qquad S^-(p, \rho) = \{x \mid x p < \rho\}$$
$$K^+(p, \rho) = \{x \mid p x = \rho\}, \qquad K^-(p, \rho) = \{x \mid x p = \rho\}.$$

For a symmetric distance we put, of course,

$$S(p, \rho) = S^+(p, \rho), \qquad K(p, \rho) = K^+(p, \rho).$$

The *length* $\lambda(x)$ of a curve $x(t)$ $(\alpha \leqq t \leqq \beta)$ is defined in the obvious way: for any partition $\Delta: \alpha = t_0 < t_1 < \cdots < t_k = \beta$ we put

$$\lambda(x, \Delta) = \sum_{i=1}^{k} x(t_{i-1})\, x(t_i) \quad \text{and} \quad \lambda(x) = \sup_{\Delta} \lambda(x, \Delta).$$

Then for each Δ

$$\lambda(x) \geqq \lambda(x, \Delta) \geqq x(\alpha)\, x(\beta) \tag{1}$$

and length has many of the usual properties, in particular, it is *additive and lower semicontinuous.* If $\lambda_{t_{i-1}}^{t_i}(x)$ is the length of $x(t) | [t_{i-1}, t_i]$ (i.e., the restriction of $x(t)$ to $[t_{i-1}, t_i]$) then

$$\lambda(x) = \sum_{i=1}^{k} \lambda_{t_{i-1}}^{t_i}(x), \quad \text{if } \alpha = t_0 < t_1 < \cdots < t_x = \beta.$$

If $x_\nu(t)$ and $x(t)$ are defined in $[\alpha, \beta]$ and $x_\nu(t) \to x(t)$ for each t then

$$\liminf \lambda(x_\nu) \geqq \lambda(x). \tag{2}$$

The curve $x(t)$ is *rectifiable* if $\lambda(x)$ is finite. We can then introduce the arc length s $(\gamma \leqq s \leqq \gamma + \lambda(x))$ as parameter by putting

$$y(s) = x(t) \quad \text{if} \quad \lambda_\alpha^t = s - \gamma.$$

A segment $T(y, z)$ from y to $z \neq y$ [1] is a curve of length yz. Because of (1) a segment $T(y, z)$ is a shortest join of y and z and any subarc of $T(y, z)$ is a segment. The term "*representation* $x(t)$ *of* $T(y, z)$" always means representation in terms of arc length and satisfies

$$(3) \qquad x(t_1)\,x(t_2) = t_2 - t_1 \quad \text{for } t_1 < t_2,$$

and any $x(t)$ ($\alpha \leqq t \leqq \beta$, $\alpha < \beta$) satisfying (3) represents a segment $T(x(\alpha), x(\beta))$.

The metric of R is *intrinsic* if xy equals the infimum of the lengths of all curves C_{xy} from x to y:

$$xy = \inf_{C_{xy}} \lambda(C_{xy}).$$

This concept presupposes that R is arcwise connected. If $T(x, y)$ exists for given $x \neq y$ then the metric is intrinsic, but the converse, in general, fails.

We need a few auxiliary facts.

(4) *If the metric of R is intrinsic then*

$$\bar{S}^+(p, \rho) = S^+(p, \rho) \cup K^+(p, \rho), \qquad \bar{S}^-(p, \rho) = S^-(p, \rho) \cup K^-(p, \rho)\,[2].$$

We must show that for $pq = \rho$ points q_ν with $pq_\nu < \rho$ and $q_\nu \to q$ exist. Choose curves C_ν from p to q with $\lambda(C_\nu) \to \rho$ and on C_ν a point q_ν with $pq_\nu = (\nu - 1)\nu^{-1}\rho$. Then (4) follows from

$$\lambda(C_\nu) \geqq pq_\nu + q_\nu q \geqq \rho.$$

(5) *If the metric is intrinsic and* $\bar{S}^+(p, \rho)$ $(\bar{S}^-(p, \rho))$ *is compact then a segment* $T(p, x)$ $(T(x, p))$ *exists for* $0 < px \leqq \rho$ $(0 < xp \leqq \rho)$. *In particular, if all* $\bar{S}^+(p, \rho)$ *are compact then* $T(p, q)$ *exists for any* $p \neq q$.

First let $0 < px < \rho$. Choose a sequence of curves C_ν from p to q with $\lambda(C_\nu) \to px$ and $\lambda(C_\nu) < \rho$. Then $C_\nu \subset \bar{S}^+(p, \rho)$ by (1) and since $\bar{S}^+(p, \rho)$ is compact, a subsequence of $\{C_\mu\}$ of $\{C_\nu\}$ converges to a curve C (l.c. (1.19)), and by (1, 2)

$$px \leqq \lambda(C) \leqq \liminf \lambda(C_\mu) = px.$$

If $px = \rho$ then (4) furnishes a sequence $\{x_\nu\}$ with $x_\nu \to x$ and $px_\nu < \rho$. Therefore, $T(p, x_\nu)$ exists and lies in $\bar{S}^+(p, \rho)$, and a subsequence of $\{T(p, x_\nu)\}$ will tend to a $T(p, x)$.

---

[1] In G, $y = z$ is admitted with $T(y, z) = y$. The restriction to $y \neq z$ proves economical.

[2] $\bar{S}^+(p, \rho)$ means the closure of $S^+(p, \rho)$. The assertion is not true without intrinsicness of the metric. If R is the subset of E^n consisting of the points x with $|x| < 1$ and $|x| \geqq 2$, then $\bar{S}(0, 2)$ does not contain $K(0, 2)$.

The following fact and its corollary are not necessary for our immediate purposes but are of general importance.

(6) *Theorem. If the metric is intrinsic,* $\bar{S}^{\pm}(p,\rho)$ *is compact,* $qp+pr\leqq\rho$ *and* $rp+pq\leqq\rho$, *then a* $T(q,r)$ *exists for* $q\neq r$.

(7) *Corollary. If* $\bar{S}^{\pm}(p,2\rho)$ *is compact and* q, r $(q\neq r)$ *lie in* $\bar{S}^{+}(p,\rho)\cap\bar{S}^{-}(p,\rho)$ *then a* $T(q,r)$ *exists.*

If $qr<\rho$ take a sequence of curves C_ν from q to r with $\lambda(C_\nu)\to qr$. If $x\in C_\nu$ and $x\notin\bar{S}^{\cup}(p,\rho)=\bar{S}^{+}(p,\rho)\cup\bar{S}^{-}(p,\rho)$ then

$$\lambda(C_\nu)\geqq qx+xr\geqq px-pq+xp-rp\geqq 2\rho-\rho=\rho.$$

Therefore, $C_\nu\subset\bar{S}^{\cup}(p,\rho)$ from a certain ν on a subsequence of $\{C_\nu\}$ tends to a $T(q,r)$.

If $qr=\rho$ then $qp+pr=\rho$. Applying (5) we obtain segments $T(q,p)$, $T(p,r)$ and $T(q,p)\cup T(p,r)$ is a $T(q,r)$.

We call

$$x(t)\quad\text{with}\quad \alpha\leqq t<\beta\quad\text{and}\quad x(t_1)\,x(t_2)=t_2-t_1\quad\text{for}\quad t_1<t_2$$

a half open segment. If for one sequence $t_\nu\to\beta$ the point $x(t_\nu)$ tends to a point b, then $\lim\limits_{t\to\beta}x(t)=b$ and defining $x(\beta)=b$ we obtain a $T(x(\alpha),x(\beta))$. We then say that the *half open segment can be completed.*

(8) *Theorem of Hopf and Rinow. In a locally compact space with an intrinsic metric the following three conditions are equivalent:*

a) *The balls* $\bar{S}^{+}(p,\rho)$ *are compact.*

b) *If* $x_\nu x_{\nu+\mu}<\varepsilon_\nu\to 0$ *for* $\nu,\mu=1,2\ldots$ *then* x_ν *converges to a point* x.

c) *Every half open segment* $x(t)$ $(\alpha\leqq t<\beta)$ *can be completed*[3].

Because each condition implies the next we prove that a) follows from c). For an arbitrary point p consider the set V of all $\rho>0$ for which $W_\rho=\bar{S}^{+}(p,\rho)$ is compact. $0<\rho'<\rho$ and $\rho\in V$ imply $\rho'\in V$. By local compactness $\rho\in V$ for small ρ. We want to show that V contains all positive ρ or that V is both open and closed in $\{\rho>0\}$.

Let W_ρ be compact and cover it with a finite number of $S^{+}(p_i,\rho_i)$ with $p_i\in W_\rho$ and compact $\bar{S}^{+}(p_i,\rho_i)$. Then $\bigcup\limits_i S^{+}(p_i,\rho_i)\supset S^{+}(p,\rho+\delta)$ for a suitable $\delta>0$, moreover $\bigcup\limits_i\bar{S}^{+}(p_i,\rho_i)\supset\bar{S}^{+}(p,\rho+\delta)$. The left side is compact, hence, also the right.

Now we show that the compactness of W_σ for $\sigma<\rho$ implies that of W_ρ or, that a sequence $\{p_\nu\}\subset W_\rho$ has an accumulation point. We

[3] In this form the theorem is due essentially to Cohn-Vossen ⟨1⟩.

may assume that $p\,p_\nu \to \rho$ and also

$$\rho_\nu = p\,p_\nu < p\,p_{\nu+1} = \rho_{\nu+1}\,^4.$$

By (5) there are segments T_ν from p to p_ν. On T_ν take the point p_ν^i with $p\,p_\nu^i = \rho_i$ $(i \leqq \nu)$. For a subsequence $\{1\nu\}$ of $\{\nu\}$ the subsegment $T(p, p_{1\nu}^1)$ of $T_{1\nu}$ converges to a segment $T_1' = T(p, q_1)$ with $p\,q_1 = \rho_1$ because W_{ρ_1} is compact. There is a subsequence $\{2\nu\}$ of $\{1\nu\}$ such that the subsegments $T(p, p_{2\nu}^2)$ of $T_{1\nu}$ converge to a segment $T_2' = T(p, q_2) \supset T_1'$ where $p\,q_2 = \rho_2$. Continuing this process we obtain segments $T(p, q_i) = T_i'$ with $p\,q_i = \rho_i$ and $T_1' \subset T_2' \subset \cdots$.

Now $\cup T_i'$ is a half open segment, hence can be completed by a point q. Then $q_i \to q$. The diagonal sequence $p_{\nu\nu}^{\nu\nu}$ also tends to q. For if $\nu \geqq i$ then $q\,p_{\nu\nu}^{\nu\nu} \leqq q\,q_i + q_i\,p_{\nu\nu}^i + p_{\nu\nu}^i\,p_{\nu\nu}^{\nu\nu} < q\,q_i + q_i\,p_{\nu\nu}^i + \rho - \rho_i$.

Since $q_i\,q \to 0$ also $q\,q_i \to 0$. Therefore i can be chosen such that $q\,q_i + \rho - \rho_i < \varepsilon/2$ and then N so large that $q_i\,p_{\nu\nu}^i < \varepsilon/2$ for $\nu > N$.

It is clear that an analogous theorem holds for the spheres $\bar{S}^-(p, \rho)$, sequences $\{x_\nu\}$ with $x_{\nu+\mu}\,x_\nu < \varepsilon_\nu \to 0$ and half open segments $x(t)$, $\alpha < t \leqq \beta$ with (3). Combining these two results we obtain

(8′) *In a locally compact space with an intrinsic metric the following conditions are equivalent:*

a) *The balls $\bar{S}^\pm(p, \rho)$ are compact.*

b) *The sequence $\{x_\nu\}$ converges if either $x_\nu\,x_{\nu+\mu} < \varepsilon_\nu \to 0$ or $x_{\nu+\mu}\,x_\nu < \varepsilon_\nu \to 0$.*

c) *Every half open segment $x(t)$ with $\alpha < t \leqq \beta$ or $\alpha \leqq t < \beta$ can be completed.*

The conditions that the $\bar{S}^+(p, \rho)$ or that the $\bar{S}^\pm(p, \rho)$ be compact distinguish the wider and the more restrictive completeness mentioned in the preface. As stated there, the theory for a symmetric distance carries over with little change to a nonsymmetric distance with compact $\bar{S}^\pm(p, \rho)$ (see Busemann [3] and Zaustinsky ⟨1, 2⟩), but the spaces with compact $\bar{S}^+(p, \rho)$ have not been studied in spite of very interesting examples. (However, see Note at the end of the text.)

We mention in particular the "*Geometrie der spezifischen Maßbestimmung*" of Funk ⟨1⟩ (see also Zaustinsky ⟨1, Appendix I⟩) where R is the interior of a closed strictly convex hypersurface H in the n-dimensional euclidean space E^n and the distance is given in terms of the euclidean distance $e(x, y)$ by $x\,x = 0$ and

$$x\,y = \log[e(x, z)/e(y, z)], \quad \text{for } x \neq y,$$

where z is the point, in which the ray from x through y intersects H.

---

[4] For, if $p\,p_\nu = \rho$ then p_ν' with $p_\nu'\,p_\nu < \nu^{-1}$ and $p\,p_\nu' < \rho$ exists by (4) and an accumulation point of $\{p_\nu'\}$ is also one of $\{p_\nu\}$.

The triangle inequality and the fact that a segment $T(x, y)$ exists and coincides (as point set) with the euclidean segment follow from the Theorem of Menelaos. If H is smooth the Finsler curvature is constant. Obviously the balls $\bar{S}^+(p, \rho)$ are compact, but the $\bar{S}^-(p, \rho)$ are not when ρ is large. As a consequence the intersection of a euclidean line with R, if not empty, has the form

$$x(t), \quad -\infty<\alpha<t<\infty \quad \text{with} \quad x(t_1)x(t_2)=t_2-t_1 \quad \text{for} \quad t_1<t_2.$$

2. Geodesic Completeness. Local Homogeneity

A *partial geodesic* in a space with a not necessarily symmetric distance is a curve $x(t)$ defined for $t \in M$, where M is a connected set (with more than one point) of the real axis $\mathbb{R}$, and each $t_0 \in M$ has a neighborhood $U(t_0)$ such that

$$x(t_1)x(t_2)=t_2-t_1 \quad \text{for} \quad t_i \in M \cap U(t_0) \quad \text{and} \quad t_1<t_2.$$

For $M=\mathbb{R}$ the partial geodesic is called a *geodesic*. A geodesic $x(t)$ is a *straight line* if $x(t_1)x(t_2)=t_2-t_1$ for any $t_1<t_2$.

The space is *geodesically complete*, if each segment can be extended to a geodesic. Precisely: if $x(t)$ $(\alpha \leqq t \leqq \beta)$ represents a segment, then a geodesic $y(t)$ $(-\infty<t<\infty)$ exists with $y(t)=x(t)$ in $[\alpha, \beta]$.

The example of Funk at the end of Section 1 shows that geodesic completeness is not a reasonable requirement unless the $\bar{S}^{\pm}(p, \rho)$ are compact.

In the literature the Hopf-Rinow Theorem is often formulated so that geodesic completeness (expressed in terms of the exponential map), or an even stronger condition discussed below, becomes one of the equivalent conditions. However, a proper closed convex subset of E^n satisfies the hypotheses of (1.8′) so that (1.8) and (1.8′) apply to interesting spaces which are not geodesically complete.

We see from this example that geodesic completeness cannot be proved without requiring the local prolongability of segments. We use the notation $(x\,y\,z)$ to indicate that $xy+yz=xz$ and that the points x, y, z are distinct ("y lies between x and z"). If a segment $T(x, y)$ exists for $x \neq y$ then prolongability is contained in (compare G (6.7)):

P: *Each point p has a neighborhood U_p such that for distinct x, y in U_p points w and z with $(w\,x\,y)$ and $(x\,y\,z)$ exist.*

(1) *Theorem. If the metric of R is intrinsic, the balls $\bar{S}^{\pm}(p, \rho)$ are compact and P holds, then R is geodesically complete.*

Each segment can be extended to a straight line if and only if P holds in the large, i.e., w and z with $(w\,x\,y)$ and $(x\,y\,z)$ exist for any $x \neq y$.

For proofs we refer to G (7.9) and (7.10). The modifications which are necessary for a nonsymmetric distance can be found in Busemann [3].

To obtain the other version of geodesic completeness we say that a partial geodesic $x(t)$ $(t \in M)$ is maximal if no partial geodesic $y(t)$ $(t \in N)$ exists where M is a proper subset of N and $y(t)=x(t)$ on M. It follows at once from Zorn's Lemma, but can also be proved without it, that

(2) *Each partial geodesic can be extended to a maximal partial geodesic.*

The second form of geodesic completeness is the *requirement that every maximal partial geodesic is a geodesic*. Since a segment is a partial geodesic it follows from (2) that this version is at least as strong as the first. Actually it is stronger.

(3) *There are ordinary metric spaces R with intrinsic metrics and compact* $\bar{S}(p, \rho)$, *where every segment can be extended to a straight line, but not every maximal partial geodesic is a geodesic.*

We show that the (x^1, x^2)-plane with the Minkowski metric $xy = \max(|x_1 - y^1|, |x^2 - y^2|)$ provides an example. Obviously every segment can be extended to a straight line (see also G (7.10)).

The convex octagon A with vertices $(\pm 1, \pm 2)$ and $(\pm 2, \pm 1)$ is a closed geodesic of length 12, because, for example, the vertices $u=(1,2)$, $v=(2,1)$, $w=(2,-1)$ have the distances $uv=1$, $vw=2$, $uw=3$. Hence $uv+vw=uw$ and the arc of A from u via v to w is a segment.

Let $x(t)$ $(-\infty<t<\infty)$ represent A, traversed infinitely often, such that $x_0=x(0)=(0,2)$, and $x(1)=v$. Then $x(t+12k)=x(t)$ for integral k. Denote by A_ε $(\varepsilon>0)$ the octagon obtained from A by the similitude

$$\bar{x}^1 = \varepsilon x^1, \quad \bar{x}^2 = \varepsilon x^2 + 2 - 2\varepsilon$$

which leaves x_0 fixed. Then A_ε is a closed geodesic of length 12ε and

$$x_\varepsilon(t) = \varepsilon x(\varepsilon^{-1} t) + (0, 2-2\varepsilon)$$

represents A_ε. Notice that $x_\varepsilon(12\varepsilon k) = x_0$.

Now let $\varepsilon_1, \varepsilon_2, \ldots$ be any positive numbers and define a partial geodesic $y(t)$ by

$$\begin{aligned} y(t) &= x(t) && \text{for } t \leqq 0 \\ &= x_{\varepsilon_1}(t) && \text{for } 0<t\leqq 12\varepsilon_1 \\ &= x_{\varepsilon_k}\left(t - 12\sum_{i=1}^{k-1}\varepsilon_i\right) && \text{for } 12\sum_{i=1}^{k-1}\varepsilon_i < t \leqq 12\sum_{i=1}^{k}\varepsilon_i, \ k>1. \end{aligned}$$

Put $\beta = 12\sum_{i=1}^{\infty}\varepsilon_i$ so that $y(t)$ is defined in $(-\infty, \beta)$. If $\beta=\infty$ then $y(t)$ is a geodesic. If β is finite, then $y(t)$ is a maximal partial geodesic (but not a geodesic). For, $y(t)\to x_0$ when $t\to\beta$. If $y(t)$ were not maximal it could be extended, whereby necessarily $y(\beta)=x_0$. But no neighborhood $U(\beta)$ exists such that

$$y(t)\,y(\beta)=\beta-t \quad \text{for} \quad t\in U(\beta) \quad \text{and} \quad t<\beta.$$

We remark in passing that this construction also gives a *negative answer to Problem* (2) *G* p. 403, which asks (in the symmetric case) whether all geodesics are straight lines if *P* holds in the large. The closed geodesic *A* and the $y(t)$ with $\beta=\infty$ provide examples.

Szenthe ⟨1⟩ was the first to notice that such examples exist. For symmetric Minkowski spaces in an affine space of any dimension he found the following criterion which will not be proved here:

(4) *A symmetric Minkowski space with unit sphere K possesses geodesics which are not straight lines (in the metric sense) if and only if K contains noncollinear points a, b, c such that the affine segments* $(1-\theta)\,a+\theta\,b$ *and* $(1-\theta)\,b+\theta\,c$ $(0\leqq\theta\leqq 1)$ *lie on K, but the triangle* $a\,b\,c$ *does not.*

Returning to the two versions of geodesic completeness we conclude from the example that uniqueness of prolongation is necessary to make them equivalent:

U: *Let* $i=1,2$. *If* $(w_i\,x\,y)$ *and* $w_1\,x=w_2\,x$ *then* $w_1=w_2$. *If* $(x\,y\,z_i)$ *and* $y\,z_1=y\,z_2$ *then* $z_1=z_2$.

The following simple observation, see Busemann [3, (4.2)], is important:

(5) *If U holds and* $(x\,y\,z)$ *then there is at most one* $T(x,y)$ *and at most one* $T(y,z)$.

Although Theorem (8.4) in *G* is not formulated in this way one readily concludes from its proof, even for a nonsymmetric distance, that

(6) *If U holds and* $x(t)$ $(\alpha\leqq t\leqq\beta)$ *represents a segment and M is any connected set in* $\mathbb{R}$ *which contains* $[\alpha,\beta]$ *then there is at most one extension of* $x(t)$ *to M as a partial geodesic. In particular, geodesic completeness is equivalent to the condition that every maximal partial geodesic is a geodesic.*

Therefore, if we add *P* and *U* to the hypotheses of the Hopf-Rinow Theorem (1.8′), then geodesic completeness in either version becomes a fourth condition equivalent to a), b), or c). Clearly, taking this as "the" Hopf-Rinow Theorem, as is frequently done, conceals the true connections. We observe also that local compactness, intrinsicness of the metric and geodesic completeness do *not* imply finite compactness. The halfplane $x^2>0$ of the (x^1,x^2)-plane with the metric $|x^1-y^1|+|x^2-y^2|$ provides an example.

The proof that a two-dimensional *G*-space is a manifold, *G* (10.4), establishes the Axiom of Pasch which is not valid for nonsymmetric distances, as pointed out by Zaustinsky ⟨1⟩, who therefore proves that the space is locally homogeneous in the sense of Montgomery [1]. This contains the desired result on two dimensions and also answers Problem 10 in *G* p. 403 affirmatively. First we must define local homogeneity.

Let M be a set in the space R with the symmetric distance $\sigma(x, y)$ and let $h(x, t) \in M$ be defined on $M \times [0, 1]$. Then $h(x, t)$ is a family of *ε-homeomorphisms* on M if

1) $h(x, 0) = x$ for $x \in M$,
2) $x \to h(x, t)$ is a homeomorphism for fixed t,
3) $\sigma(x, h(x, t)) < \varepsilon$ for $x \in M$ and $0 \leqq t \leqq 1$.

The space R is *locally homogeneous* if every point has a neighborhood U with the following properties: *Given* $\varepsilon > 0$ *there is a* $\delta > 0$ *such that for* $a \in \bar{U}$ *and* $\sigma(a, b) < \delta$ (b need not be in $\bar{U}$) *an ε-family of homeomorphisms* $h(x, t)$ *on* $\bar{U}$ *with* $h(a, 1) = b$ *exists.*

(7) *Theorem. A space R with an intrinsic distance and compact* $\bar{S}^+(p, \rho)$ [5] *which satisfies the conditions P and U is locally homogeneous.*

We introduce the symmetric distance $\sigma(x, y) = \max(x\,y, y\,x)$ and put

$$S(p, \rho) = \{x \mid \sigma(p, x) < \rho\} = S^+(p, \rho) \cap S^-(p, \rho).$$

Then $\bar{S}(p, \rho)$ is compact. For a given point p and a given $\varepsilon > 0$ we choose $0 < \rho_1 < \rho_2 < \rho_3 < \rho_4 < \varepsilon/2$ such that with $S(p, \rho_i) = S_i$ the ball $\bar{S}_4$ lies in a neighborhood in which P holds, $T(x, y) \subset \bar{S}_4$ if x, y lie in $\bar{S}_3$, and $T(a, b) \subset \bar{S}_3$ if a, b lie in S_2.

We prove local homogeneity with the given ε for $a \in \bar{S}_1$ and $\delta = \rho_2 - \rho_1$. The assertion is trivial for $a = b$. So let $a \neq b$ and $b \in S_2$ (which is true if $a\,b < \delta$). Since $T(a, b) \subset S_3$ there is by P a $T(a, c) \supset T(a, b)$ with $c \in \bar{S}_3 - S_3$ and $T(a, c) \subset S_3 \cup \{c\}$. (The argument is the same as for *G* (7.8).)

For $x \in \bar{S}_1$ the segment $T(x, c)$ is unique and lies in $\bar{S}_4$. Therefore, if $z_x(t)$ represents $T(x, c)$ with $z_x(0) = x$, then $z_x(t)$ is continuous in x and t. With $\beta = a\,b / a\,c$ put

$$h(x, t) = z_x(t\,\beta \cdot x\,c), \qquad 0 \leqq t \leqq 1.$$

Then $h(x, 0) = z_x(0) = x$, $h(a, 1) = z_a(a, b) = b$, and $h(x, t)$ is continuous in $\bar{S}_1 \times [0, 1]$.

Because $\bar{S}_1$ is compact, it only remains to show that

$$y = h(x_1, t) = h(x_2, t) \quad \text{implies } x_1 = x_2.$$

This is obvious for $t = 0$, so let $t > 0$. Then $(x_i\,y\,c)$. If $x_1\,y = x_2\,y$ then uniqueness of prolongation gives $x_1 = x_2$. We show that, for example, $x_1\,y < x_2\,y$ is impossible. In that case $(x_2\,x_1\,y)$ and $(x_2\,x_1\,c)$ because an x_1^1

[5] The hypotheses can be weakened, but this is the most significant case. All that is needed is that R be locally compact.

on $T(x_2, y)$ with $x_2 x_1 = x_2 x_1^1$ exists. Then $(x_2 x_1^1 y)$ gives $x_1 = x_1^1$ since $T(x_2, y)$ is unique. But then $y = z_{x_1}(t\beta \cdot x_i c)$ would yield

$$x_2 x_1 = x_2 y - x_1 y = t\beta(x_2 c - x_1 c) = t\beta x_2 x_1 < x_2 x_1.$$

The importance of local homogeneity lies in the fact that a finite dimensional separable space satisfying this condition has many properties of manifolds. Since $R = \bigcup_{k=1}^{\infty} \bar{S}^+(p, k)$ the spaces with compact $\bar{S}^+(p, \rho)$ are separable. We conclude from Montgomery [1].

(8) *Theorem. Let R satisfy the hypotheses of Theorem* (7). *If R has dimension* 1 *or* 2 *then it is a manifold.*

If $\dim R = n < \infty$ *then any two points have homeomorphic neighborhoods. A closed n-dimensional set has interior points. Any open set* $V \neq \emptyset$ *contains an open set* $W \neq \emptyset$ *such that any* $(n-1)$*-cycle in W bounds in V.*

Before leaving these considerations we emphasize that deep and unexpected theorems of differential geometry in the large can be proved for spaces with compact $\bar{S}^+(p, \rho)$ and intrinsic metrics even without P and U. Section 12 contains an outstanding example: Under these weak conditions we will there extend to n dimensions a theorem which even in the Riemannian case was known only for two.

3. The Topology of *r*-Spaces

This section deals with the purely topological theory of r-spaces by Kosiński [1], which we quote as [*K*]. It is relevant for us because G-spaces are r-spaces, see Section 4. The principal conclusions are that (i) a finite dimensional G-space has the property of domain invariance and (ii) small spheres $K(p, \rho)$ are not contractible, i.e., cannot be deformed over themselves to a point (see Dugundji, ⟨1, p. 316⟩).

For brevity, we will call an open set $V \neq \emptyset$ in a metric space *canonical* if $\bar{V}$ is compact and for each $p \in V$ the boundary $\dot{V}$ of V is a strong deformation retract of $\bar{V}$-$\{p\}$; this means by definition (l.c. p. 324) that $\bar{V}$-$\{p\}$ can be deformed over itself into $\dot{V}$ keeping each point of $\dot{V}$ fixed during the entire deformation.

An r-space is a metric space of constant finite dimension $n \geqq 1$, where each point has arbitrarily small canonical neighborhoods; since the canonical neighborhoods are relatively compact, each r-space is therefore locally compact. This definition differs from that of [*K*] in two respects. The first is minor: [*K*] does not require constant dimension but deduces this from connectedness which occurs as a hypothesis in all theorems that concern us here. G-spaces have constant dimension, see (2.8) or G (10.2). The second is important: [K] assumes compactness of the space and uses it in an essential way. However, this assumption is not appropriate for G-spaces; and reduction of the non-compact to the compact case is not immediately accomplished by compactification[6].

Nevertheless, *the results of Kosiński in which we are interested are valid for the locally compact case*; to ascertain this, we will use a homology theory slightly different from that in [K]. We shall derive the required results by using Cech homology on compact carriers with the reals mod 1 as coefficients[7]; as is well-known, ⟨Eilenberg-Steenrod 1 (quoted as E-S), p. 271⟩ this theory reduces to the classical Cech theory on compact pairs in the given locally compact space. The space is called R and we assume throughout this section that R is an r-space; we use $A \subset\subset B$ to denote that A is a proper subset of B.

(a) *Let* $\dim R \leqq n$, *and let W be a canonical set. Then* $H_n(B \cup \dot{W}, \dot{W}) = 0$ *for all closed* $B \subset\subset \overline{W}$.

Clearly, $W \not\subset B$ since otherwise $\overline{W} \subset \overline{B} = B$; therefore a $p \in W - B$ exists. Now consider the exact sequence

$$\cdots \to H_{n+1}(\overline{W} - p, B \cup \dot{W}) \to H_n(B \cup \dot{W}, \dot{W}) \to H_n(\overline{W} - p, \dot{W}) \to \cdots$$

Since $\dim \overline{U} \leqq n$ and we are using Cech homology on compact carriers over the reals mod 1, it follows immediately from [Hurewicz-Wallman 1 (quoted as H-W), Theorem VIII 4, p. 152] that $H_{n+1}(\overline{W} - p, B \cup \dot{W}) = 0$; furthermore, $H_n(\overline{W} - p, \dot{W}) = 0$ because $\dot{W}$ is a strong deformation retract of $\overline{W} - p$. By extractness, we conclude that $H_n(B \cup \dot{W}, \dot{W}) = 0$.

(b) *Let* $\dim R \leqq n$, *and let W be any canonical set. Then* $H_n(B) = 0$ *for all closed* $B \subset\subset W$.

Since $B \cap \dot{W} = \emptyset$, we find first that $H_n(B) = H_n(B, B \cap \dot{W})$; then by excision [$E$-$S$, p. 207] and (a) that $H_n(B, B \cap \dot{W}) = H_n(B \cup \dot{W}, \dot{W}) = 0$.

(c) *If W is a canonical set and* $\dim R = n$, *then* $H_n(\overline{W}, \dot{W}) \neq 0$.

Proof. Choose a canonical V with $\overline{V} \subset W$; since $\overline{V}$ is compact and $\dim \overline{V} = n$, there is by [$H-W$, p. 152], a closed $A \subset \overline{V}$ such that $H_n(\overline{V}, A) \neq 0$. Because of (b), we have $H_n(\overline{V})$ so, from the exact sequence

$$\cdots \to H_n(\overline{V}) \to H_n(\overline{V}, A) \xrightarrow{\partial} H_{n-1}(A) \to \cdots$$

we conclude that ∂ is monic and therefore that $H_{n-1}(A) \neq 0$; moreover $A \cap \dot{W} = \emptyset$ so by excision we find that $H_{n-1}(A \cup \dot{W}, \dot{W}) = H_{n-1}(A) \neq 0$.

[6] This can be seen from the fact that a small finite-dimensional ball $\bar{S}(p, \rho)$, $\rho > 0$, in a G-space is not an r-space; see the corollary in [K], p. 118 which states that for compact r-spaces R, R' of the same finite dimension, $R \supset R'$ implies $R = R'$.

[7] I am indebted to J. Dugundji for much of the version that follows.

Choose a non-zero element $\alpha \in H_{n-1}(A \cup \dot{W}, \dot{W})$, which we keep fixed for the remainder of the proof. For any $Q \supset A \cup \dot{W}$, we let $\alpha(Q)$ be the image of α in the homomorphism $i: H_{n-1}(A \cup \dot{W}, \dot{W}) \to H_{n-1}(Q, \dot{W})$ induced by the inclusion map.

Now let $p \in W - A$; then $\alpha(\overline{W} - p) = 0$, since $H_{n-1}(\overline{W} - p, \dot{W}) = 0$; moreover, there is a neighborhood U of p such that $\alpha(\overline{W} - U) = 0$: one need only take U to be a neighborhood of p that does not meet the (compact) trace described by A during the deformation retraction of $\overline{W} - p$ onto $\dot{W}$.

Noting that $\overline{W} - U$ is compact, the continuity of the Cech theory on compact pairs [$E - S$, p. 261] together with a simple application of the Brouwer reduction theorem [$H - W$, p. 161] shows that there exists a closed $P \subset \overline{W}$ such that

(i) $A \cup \dot{W} \subset P \subset \overline{W} - U$,

(ii) $\alpha(P) = 0$, but $\alpha(L) \neq 0$ for any closed L, $A \cup \dot{W} \subset L \subset\subset P$. It is evident that $P \neq A \cup \dot{W}$, since $\alpha(A \cup \dot{W}) = \alpha \neq 0$, so there exists a point $q \in P \cap (W - A)$. Repeating the above argument with q instead of P, there exists a neighborhood U_1 of q and a closed $F \subset \overline{W}$ that

(iii) $A \cup \dot{W} \subset F \subset \overline{W} - U_1$,

(iv) $\alpha(F) = 0$, but $\alpha(L) \neq 0$ for any closed L, $A \cup \dot{W} \subset L \subset\subset F$. It is clear that $F \neq P$.

Now consider the Mayer-Vietoris exact sequence ⟨Spanier 1, p. 209⟩

$$\cdots \to H_n(P \cup F, \dot{W}) \xrightarrow{\partial} H_{n-1}(P \cup F, \dot{W}) \xrightarrow{i} H_{n-1}(P, \dot{W}) \oplus H_{n-1}(F, \dot{W}) \to \cdots$$

where $i = i_1 \oplus i_2$, the homomorphisms induced by inclusion. The element $\alpha(P \cap F) \in H_{n-1}(P \cap F, \dot{W})$ is not zero, because of $P \cap F \subset\subset P$ and (ii). However $i_1[\alpha(P \cap F)] = \alpha(P) = 0$ and $i_2[\alpha(P \cap F)] = \alpha(F) = 0$ so, by exactness, there must exist a (necessarily non-zero) element β in $H_n(P \cup F, \dot{W})$ with $\partial\beta = \alpha(P \cap F)$. Thus, $H_n(P \cup F, \dot{W}) \neq 0$ and, since $(P \cup F) = (P \cup F) \cup \dot{W}$, we find from (a) that $P \cup F = \overline{W}$. This completes the proof.

(d) *Let* $\dim R = n$, *and let* W *be any canonical set. Then* $H_n(\overline{W}, \overline{W} - G) \neq 0$ *for each non-empty open* $G \subset W$.

Consider the exact sequence

$$\to H_n(\overline{W} - G, \dot{W}) \to H_n(\overline{W}, \dot{W}) \xrightarrow{i} H_n(\overline{W}, \overline{W} - G) \to \cdots.$$

We have $\overline{W} - G \subset\subset \overline{W}$ and $(\overline{W} - G) = (\overline{W} - G) \cup \dot{W}$ so, by (a), $H_n(\overline{W} - G, \dot{W}) = 0$. Thus, i is monic and, since $H_n(\overline{W}, \dot{W}) \neq 0$, so also $H_n(\overline{W}, \overline{W} - G) \neq 0$.

(e) *Let* $\dim R = n$, *let* W *be canonical, and let* G *be any non-empty open set with* $\bar{G} \subset W$. *Then*

$$(e_1) \qquad H_n(\bar{G}, \dot{G}) \neq 0$$

$$(e_2) \qquad 0 \to H_n(\bar{G}, \dot{G}) \xrightarrow{\partial} H_{n-1}(\dot{G}) \xrightarrow{i} H_{n-1}(\bar{G})$$

is exact.

Proof. To prove (e_1), we first observe that the inclusion $i: (\bar{G}, \dot{G}) \to (\bar{W}, \bar{W} - G)$ is a relative homeomorphism of compact pairs, so that [$E-S$, p. 266] $H_n(\bar{G}, \dot{G}) = H_n(\bar{W}, \bar{W} - G)$ and then apply (d). Exactness of the sequence in (e_2) follows from (b).

Statements (d) and (e) are the decisive facts in the theory. In the first place, (e_1) and (e_2) imply that $H_{n-1}(\dot{G}) \neq 0$; in particular, we have

(1) *Theorem. If* $\dim R = n$ *and* W, V *are canonical sets with* $\bar{V} \subset W$, *then* $\dot{V}$ *is not contractible.*

Secondly, as done in [K], the statements permit a characterization of the interior points of a compact subset $E \subset R$ in terms of the topology of E alone entirely analogous to that used in euclidean spaces. To formulate it, we denote by $V_E, U_E, \ldots$ sets open in the relative topology of E. Each $V_E = V \cap E$ for some set V open in R; and whenever E is compact, the closures of V_E in E and in R coincide.

(f) *Let* $\dim R = n$, *and let* $E \subset R$ *be any compact set. Then* $p \in E$ *is an interior point of* E *if and only if for each neighborhood* U_E *of* p, *there exists a neighborhood* V_E *of* p, $\bar{V}_E \subset U_E$, *with the property:* $H_n(\bar{V}_E, \bar{V}_E - G_E) \neq 0$ *for every neighborhood* G_E *of* p *with* $\bar{G}_E \subset V_E$.

Proof. Assume that p is an interior point of E. Then in each neighborhood U_E of p, there is a canonical open (in R) set V with $p \in V \subset \bar{V} \subset U_E$; according to (d), the neighborhood $V_E = V \cap E = V$ has the required property.

Now let $p \in E$ be a boundary point; it is enough to show that for each given neighborhood $V_E = V \cap E$ of p, we can find a neighborhood G_E of p, $\bar{G}_E \subset V_E$, such that $H_n(\bar{V}_E, \bar{V}_E - G_E) = 0$. Select any canonical G, $p \in G \subset \bar{G} \subset V$; since E is compact and p is a boundary point, we have $\bar{G}_E = E \cap \bar{G}$ is a proper closed subset of $\bar{G}$.

Let $i: (\bar{G}_E, \bar{G}_E \cap \dot{G}) \to (\bar{V}_E, \bar{V}_E - G_E)$ be the inclusion map; this is evidently a relative homeomorphism of compact pairs, so [$E-S$, p. 266] that

$$H_n(\bar{V}_E, \bar{V}_E - G_E) = H_n(\bar{G}_E, \bar{G}_E \cap \dot{G}).$$

Each set of the pair $(\bar{G}_E, \bar{G}_E \cap \dot{G})$ is contained in the compact $\dot{G}$; by excision,

$$H_n(G_E, \bar{G}_E - \dot{G}) = H_n(\bar{G}_E \cup \dot{G}, \dot{G}).$$

However, $\bar{G}_E \subset\subset \bar{G}$ so, by (a), we find $H_n(\bar{G}_E \cup \dot{G}, \dot{G})=0$. This completes the proof.

The principal application of (f) is

(2) *Theorem. If R, R' are r-spaces of the same finite dimension n, and if ϕ maps the open set $D \subset R$ topologically into R', then ϕD is open in R'.*

Let $p' \in \phi D$ and $p=\phi^{-1}p'$. Take canonical neighborhoods V, W with $p \in V \subset \bar{V} \subset W \subset D$; then $\phi\bar{V}$ is compact, and $p' \in \phi\bar{V} \subset \phi D$. Since p is an interior point of $\bar{V}$, we can apply (f) to $\bar{V}$ as E. The conditions for an interior point in (f) are therefore satisfied by p and these are invariant under ϕ, so they hold also for p' and $\phi\bar{V}$. Thus, p' is an interior point of $\phi\bar{V}$, and therefore also of ϕD.

We say that *domain invariance* holds in the space R whenever R has the property: if one of two homeomorphic subsets of R is open in R, then so is the other. Theorem (2) implies

(3) *Corollary. Domain invariance holds in any finite-dimensional r-space.*

A further consequence of our considerations is

(4) *Let V, W be canonical sets in an n-dimensional r-space, and let $\bar{V} \subset W$. Then there is no continuous map of $\bar{V}$ into a proper subset of itself that leaves $\dot{V}$ pointwise fixed.*

Proof. Assume that such a map, f, did exist; then $\dot{V} \subset f(\bar{V}) \subset\subset \bar{V}$ so that a $p \in V - f(\bar{V})$ exists. Let $r: \bar{V}-p \to \dot{V}$ be a strong deformation retraction; then $r \circ f$ shows that $\dot{V}$ is a retract of $\bar{V}$and therefore [Eilenberg-Steenrod, p. 50] that

$$H_n(\bar{V}) = H_n(\dot{V}) \oplus H_n(\bar{V}, \dot{V}).$$

This is impossible, since $H_n(\bar{V})=0$ by (b) whereas $H_n(\bar{V}, \dot{V}) \neq 0$ by (c).

We remark that, as our proof of (f) shows (and that given for an essentially equivalent statement in [K] does not) the assertion (f), and therefore domain invariance, follows simply from the existence of a homology theory in a suitable category of pairs that has the properties (a) and (c) for some $n \neq 0$.

4. Finite-Dimensional *G*-Spaces

With few exceptions we will from now on only consider G-spaces. The axioms I to V for G-space R are listed in G, p. 37. They amount to the following:

R is an ordinary metric space with compact $\bar{S}(p, \rho)$ and an intrinsic metric which satisfies the axioms P of local prolongability and U of uniqueness of prolongation (see Section 2)[8].

[8] P and U are the axioms IV and V. The compactness of the balls $\bar{S}(p, \rho)$ is the finite compactness in II and the equivalence of III with intrinsicness of the metric follows from (1.5).

Equivalent conditions are obtained from the Hopf-Rinow Theorem and the remark following (2.6).

Two important functions $\rho(p)$ and $\rho_1(p)$ are defined in G, pp. 33, 39, namely the suprema of those ρ for which P holds with $U_p=S(p,\rho)$, respectively $T(x,y)$ is unique for x,y in $S(p,\rho)$. Since $T(x,y)$ is unique when z with (xyz) exists, we have $\rho(p)\leqq\rho_1(p)$. The question whether $\rho(p)=\rho_1(p)$ is discussed below. $\rho(p)$ has the property:

(1) *Either* $\rho(p)\equiv\infty$ *or* $0<\rho(p)<\infty$ *and* $|\rho(p)-\rho(q)|\leqq pq$ *and the same holds for* $\rho_1(p)$ (G, pp. 33, 39).

The question (G, p. 403, Problem (7)) whether every G-space has a finite dimension remains open.

It is proved in G and also follows from (2.8) that one- and two-dimensional G-spaces are manifolds. The question whether a finite-dimensional G-space is a manifold was called inaccessible in G, p. 403 and this proved, unfortunately, correct. There is also conjectured (Problem (5)), that the three-dimensional case might prove accessible. In fact, Krakus ⟨1⟩ proved recently:

(2) *Three-dimensional G-spaces are topological manifolds.*

The case where small spheres are convex is also contained in Rolfsen ⟨1⟩. Krakus showed:

If R is an at least three-dimensional G-space, then any sphere $K=K(p,\rho)$ with $0<\rho<\rho(p)$ is arcwise and locally arcwise connected, cyclic, unicoherent, and each point of K has arbitrarily small neighborhoods U (on K) such that $K-U$ is acyclic.

It then follows from an early theorem of Borsuk ⟨2⟩, of which the author was not aware, that K is homeomorphic to S^2 if $\dim K=2$ or $\dim R=3$. Obviously $S(p,\rho)$ is then homeomorphic to E^3 and R is a manifold.

That K is arcwise and locally arcwise connected is shown in G (55.1). The proof of (9) below provides for any two antipodal points w, w' of K a contraction x_t $(0\leqq t\leqq 1)$ of $K-\{w'\}$ into w. This yields the obvious fact that K is cyclic, i.e., $K-\{w'\}$ is connected, and also the existence of small neighborhoods U' of w' with acyclic $K-U'$. For, if U is any neighborhood of w' form $F_u=\{x_t|x\in\overline{K-u},\,0\leqq t\leqq 1\}$.

Then F_u is closed and contains $\overline{K-u}$ but not w', hence $w'\in U'=K-F_u<U$ and x_t defines a contraction of $K-U'$ into w, so that $K-U'$ is acyclic.

Finally, if K were not unicoherent (Dugundji ⟨1, p. 364⟩) then a Jordan curve J and a retraction ψ of K onto J would exist, see Borsuk ⟨1, p. 184⟩, and J is a proper subset of K because $\dim K\geqq 2$. Then (9) furnishes a contractible C with $J\subset C\subset K$. If ϕ: $C\times[0,1]\to K$ is a contraction of C, then $\psi\circ\phi|J\times[0,1]$ would yield a contraction of J over itself.

To apply the results of the last section we prove, remembering that $\bar{S}(p,\rho)=S(p,\rho)\cup K(p,\rho)$:

(3) *Theorem. If* $0<\rho<\rho(p)$ *then for any* $q\in S(p,\rho)$ *the sphere* $K(p,\rho)$ *is a strong deformation retract of* $\bar{S}(p,\rho)-\{q\}$. *Hence G-spaces are r-spaces.*

Because the balls $\bar{S}(p,\rho)$ (spheres $K(p,\rho)$) are homeomorphic for $0<\rho<\rho(p)$ we may assume that $0<\rho<\rho(p)/4$. Put $\bar{S}=\bar{S}(p,\rho)$, $K=K(p,\rho)$.

The assertion is trivial for $q=p$ (and is trivial for all q if $\bar{S}$ is convex, i.e., contains $T(x,y)$ with x and y). Let $q\neq p$. With $2\varepsilon=\rho-pq$ consider the "ellipsoid"

$$E=\{x|px+xq=pq+\varepsilon\},\qquad \text{also } E_i=\{x|px+xq<pq+\varepsilon\}.$$

Then

$$(\alpha)\ E\cup E_i\subset S,\qquad (\beta)\ S(p,\varepsilon/2)\subset E_i,\qquad (\gamma)\ S(q,3\rho(q)/4)\supset E\cup E_i.$$

For, if $x\in E\cup E_i$ then $px<pq+\varepsilon\leqq\rho-\varepsilon$ proving (α). Next, $px+xq\leqq 2px+pq<\varepsilon+pq$ for $px<\varepsilon/2$, whence (β). If $qr\geqq 3\rho(q)/4$ then (γ) follows from $pr\geqq 3(\rho(q)-pq)/4\geqq 2\rho>pq+\varepsilon$, so $r\notin E\cup E_i$.

If x traverses $T(q,r)$ with $qr=3\rho(q)/4$ from q to r then $qx+xp$ increases from $pq<pq+\varepsilon$ to $qr+rp>pq+\varepsilon$, so $T(q,r)$ intersects E. The intersection is unique, because $qx+xp$ increases on $T(q,r)$ except for $p\in T(q,r)$ and $x\in T(q,p)$ when $px+xq=pq<pq+\varepsilon$. For, (qxy) implies $px+xq<py+yx+xq=py+yq$ unless x,y lie on $T(p,q)$.

Therefore, $\bar{E}_i=E\cup E_i$. Clearly we can now deform $\bar{E}_i-\{q\}$ over itself leaving E pointwise fixed by sliding every point on a $T(q,r_E)$ with $r_E\in E$. We extend this deformation to $\bar{S}$ by letting every point of $\bar{S}-\bar{E}_i$ stay fixed during the deformation. Because of (β) we can now deform $\bar{S}-\bar{E}_i$ radially from p over $\bar{S}-\bar{E}_i$ into K leaving K pointwise fixed.

We conclude from (3), (3.1, 3):

(4) *Theorem. In finite-dimensional G-spaces domain invariance holds and spheres* $K(p,\rho)$ *with* $0<\rho<\rho(p)$ *are not contractible.*

The question, whether the lack of topological information regarding G-spaces is not very disturbing, is frequently raised. Actually the two properties occurring in (4) are the only ones which were hitherto encountered and that only in very few places. Therefore finite dimensionality suffices for all purposes.

Domain invariance enters into a single, but important theorem of G, namely (38.2). Using (4) we conclude from (38.2):

(5) *The universal covering space of a G-space with finite dimension and convex capsules (hence also with nonpositive curvature) is straight (i.e., all geodesics are straight lines).*

Domain invariance will occur here in Sections 9 and 11. Notice also the following completeness property:

(6) *A proper subset E of a G-space R which contains interior points is with the metric of R not a G-space. If* $\dim R = n$ *it suffices that* $\dim E = n$.

For the first part let p be an interior point of E and $q \in R - E$. Consider a segment T from p to q. If $T \cap E$ is not closed then E is not complete. If $T \cap E$ is closed then there is a last point r on T such that the subsegment $T(p, r)$ of T lies in E and $r \neq p, q$. But G (7.8) does not hold at r in E.

In the second part we apply Theorem (3.2) to the inclusion ϕ of E (as R) to R (as R') and find that E consists of interior points of R which contradicts the first part.

An immediate consequence of (3) and (3.4) is:

(7) *In a finite-dimensional G-space there is no continuous map of* $\bar{S}(p, \delta)$ $(0 < \delta < \rho(p))$ *on a proper subset which leaves* $K(p, \delta)$ *pointwise fixed.*

This yields an analogue to a well known theorem. Call the set C in a G-space strictly convex if for $x \neq y$ in C the segment $T(x, y)$ is unique and $T(x, y) - \{x \cup y\}$ lies in the interior of C. (Hence C has interior points unless it consists of one point.)

(8) In a finite-dimensional G-space, if some ball $S(p, \delta)$ with $0 < \delta \leqq \rho(p)/4$ is strictly convex, then any continuous map ψ of $\bar{S}(p, \rho)$ $(0 < \rho < \rho(p))$ into itself has a fixed point.

Assume ψ has no fixed point. Then the obvious homeomorphism of $\bar{S}(p, \rho)$ on $\bar{S}(p, \delta)$ produces a map ϕ of $\bar{S}(p, \delta)$ on itself without fixed points. Since $\phi x \neq x$ there is a unique $T(\phi x, y_x)$ containing x and of length $3\rho(p)/4$, because $\phi x x \leqq \rho(p)/2$ and

$$\rho(\phi x) \geqq \rho(p) - p\phi x \geqq 3\rho(p)/4.$$

Moreover, $T(\phi x, y_x)$ intersects $K(p, \delta)$ in a point z_x, since

$$p y_x \geqq \phi x y_x - p \phi x \geqq 3\rho(p)/4 - \rho(p)/4 = \rho(p)/2$$

and z_x is unique because $\bar{S}(p, \delta)$ is strictly convex. Therefore z_x depends continuously on x and $x \to z_x$ would map $\bar{S}(p, \delta)$ on $K(p, \delta)$ leaving $K(p, \delta)$ pointwise fixed in contradiction to (7).

We now turn to facts which hinge on the hypothesis that $K(p, \rho)$ with $0 < \rho < \rho(p)$ is not contractible and are therefore valid in finite-dimensional G-spaces. They all concern the existence of prolongations and are based on an idea of A.D. Alexandrov ⟨1⟩. Two lemmas are needed.

(9) *If w, w' are antipodal points on* $K = K(p, \delta)$ $(0 < \delta < \rho(p))$ *then* $K - \{w'\}$ *can be contracted into w.*

For $x \in K - \{w'\}$ let y be the point with (pyx) and $py = \delta/4$, in particular let (psw) and $ps = \delta/4$. Then $T(s, y)$ lies in $S(p, \delta)$ (see G (6.9)) does not pass through p and is unique. Let y_t be the point of $T(s, y)$ with $sy_t = (1-t)\,sy$ $(0 \leqq t \leqq 1)$. Then $y_t \neq p$ and x_t with $(py_t x_t)$ and $x_t \in K - \{w'\}$ exists, because $x_t = w'$ would imply (wpy) and $x = w'$. Because $T(s, y)$ is unique the point x_t depends continuously on x and t, moreover $x_0 = x$ and $x_1 = w$ for all $x \in K - \{w'\}$.

(10) *Let U be an open set (in a G-space) with the following property: there are disjoint subsets M and N of U such that for $x \in M$ and $y \in N$ a unique shortest connection $C(x, y)$ of x and y in U exists. Then $C(x, y)$ is a geodesic arc and depends continuously on x and y $((x, y) \in M \times N)$.*

The first assertion is obvious: Because $C(x, y)$ is a shortest connection it is an arc and locally a segment, hence a geodesic arc.

Let $x_\nu, x \in M$ and $x_\nu \to x$, $y_\nu, y \in N$ and $y_\nu \to y$. We must show that

$$C_\nu = C(x_\nu, y_\nu) \to C(x, y) = C.$$

Because U is open $T(x_\nu, x)$ and $T(y_\nu, y)$ lie in U for large ν. Therefore the lengths λ_ν of C_ν and λ of C satisfy

$$|\lambda_\nu - \lambda| \leqq x_\nu x + y_\nu y, \qquad \text{so } \lambda_\nu \to \lambda.$$

If C_ν did not converge to C a subsequence $\{C_\mu\}$ would tend to a curve $C_0 \neq C$ with length

$$\lambda_0 \leqq \liminf \lambda_\mu = \lambda.$$

The uniqueness of $C(x, y)$ guarantees that $C_0 \not\subset U$. However, C_0 must begin and end as a segment (as limit of C_μ). Therefore $z \in U \cap C_0$ exists which decomposes C_0 into C_0' and a $T(z, y)$. On C_μ there is a $z_\mu \to z$ and if z is close to y the point z_μ decomposes C_μ into an arc C_μ' and a $T(z_\mu, y_\mu)$.

The arc $C(x, z)$ cannot have $T(y, z)$ as a prolongation, because prolonging $T(y, z)$ beyond z leads outside U. Therefore

$$\lambda(C(x, z)) + zy = \lambda + 2\eta \qquad \text{with } \eta > 0.$$

But

$$\lambda_\mu = \lambda(C_\mu') + z_\mu y_\mu \to \lambda$$

and

$$\lambda(C(x, z)) > \lambda(C_\mu') + \eta \qquad \text{for large } \mu,$$

which contradicts $|\lambda(C(x, z)) - \lambda(C_\mu')| \leqq z z_\mu$.

We apply (9) and (10) in the proof of

(11) *Theorem. In the open set U of a G-space let two distinct points p, q and $S(q, \rho_0) \subset U$ exist such that for $x \in S(q, \rho_0)$ there is a unique shortest connection $C(p, x)$ of p and x in U. If the spheres $K(q, \rho)$ $(0 < \rho < \rho(p))$ are not contractible (hence if* $\dim R < \infty$*) then q lies on some $C(p, x)$ with $x \neq q$.*

The point is here, of course, that $C(p,q)$ can be extended beyond q so that it remains a unique shortest connection of its end points in U. Notice that (11) strongly resembles a well known theorem on conjugate points in the calculus of variations.

Let $0<\delta<\min(\rho_0, \rho(q), pq)$. Because $C(p,q)$ is a shortest connection it intersects $K=K(q,\delta)$ in exactly one point w. Denote the antipode of w on K by w'. If (puq) and $uq=\delta/2$ then the projection P of $S(u,\delta/4)$ from q on K by segments from q is a proper subset of K, in particular $w'\notin P$.

For $0<\varepsilon<\delta$ put $K_\varepsilon=K(q,\varepsilon)$ and form

$$V=\bigcup_{v\in K_\varepsilon} C(p,v).$$

It suffices to prove that $q\in V$ for a suitable ε since $q\notin K_\varepsilon$. Let $z(t,v)$ represent $C(p,v)$ with $z(0)=v$. Since $uq=\delta/2$ we can choose ε so small that $z(\delta/2,v)\in S(u,\delta/4)$ and $T(z(\delta/2,v),v)\subset S(q,\delta)$, using $T(z(\delta/2,v)\subset C(p,v)$ and (10). If $z(t\delta/2,v)\neq q$ $(0\leqq t\leqq 1)$ we define y_t by $(qz(t\delta/2,v)\,y_t)$ and $qy_t=\delta$. For $t=0$ we have (qvy_0) and y_0 traverses K because v traverses K_ε. The point y_1 lies in P. By (10) the point $z(t\delta/2,v)$ depends continuously on t and v. Therefore y_t does too, so that y_t would define a deformation of K into a proper subset M which does not contain w'. Using (9) we could then deform M into w.

This has the following important implications.

(12) *Theorem. In a G-space in which the* $K(p,\rho)$ $(0<\rho<\rho(p))$ *are not contractible, hence in every finite-dimensional G-space, the following hold:*

(a) *If* $T(p,x)$ *is unique for* $x\in S(q,\rho)$ $(q\neq p, \rho>0)$ *then* r *with* (pqr) *exists,*

(b) $\rho(p)\equiv\rho_1(p)$.

(c) *If* $T(x,y)$ *is unique for any* $x\neq y$ *then the space is straight.*

(d) *Let* $V_1(p)$ *consist of* p *and those points* x *for which* $T(p,x)$ *is unique and* $V(p)$ *of* p *and those* x *for which* y *with* (pxy) *exists. Then* $V(p)\subset V_1(p)$ *and*

$$\operatorname{Int} V(p)=\operatorname{Int} V_1(p).$$

(e) *If* $\delta_1(p)=\sup\rho$ *where* $T(p,x)$ *is unique for* $px<\rho$ *and* $\delta(p)=\sup\rho$ *where* (pxy) *exists for* $px<\rho$ *then* $\delta(p)=\delta_1(p)$.

(f) *If* $T(p,x)$ *is unique for every* $x\neq p$ *then each* $T(p,x)$ *can be extended to a ray with origin* p.

(a) is an immediate corollary of (11) applied to $U=R$. To prove (b) we remember that $\rho(p)\leqq\rho_1(p)$ holds in any G-space. If x,y are distinct points in $S(p,\rho_1(p))$ then $U=S(y,\rho_1(p)-py)\subset S(p,\rho_1(p))$ and $T(x,v)$ is unique for $v \in U$. By (a) a point z with (xyz) exists, which means by the definition of $\rho(p)$ that $\rho(p) \geqq \rho_1(p)$.

In particular $\rho_1(p) \equiv \infty$ implies $\rho(p) \equiv \infty$ and this means that the space is straight. Problem (3) G p. 403 asks whether this is true for all G-spaces. In this generality the question remains open.

(d), (e), and (f) are immediate consequences of (a).

It was mentioned that the idea of the proof of (11) stems from Alexandrov ⟨1⟩. He establishes the following version of (12(a)), compare also Rinow ⟨1, pp. 288, 289⟩:

(13) *Let p and q be distinct points in a finitely compact metric space and let q have a neighborhood U homeomorphic to E^n such that $T(p, x)$ exists for $x \in U$ and is unique. Then there is an $r \subset U - \{q\}$ with $T(p, r) \supset T(p, q)$.*

In the proof one introduces an auxiliary euclidean metric $e(x, y)$ in U and operates with the euclidean spheres $\{x \mid e(p, x) = \rho\}$ instead of on $K(p, \rho)$, also, one uses the euclidean projection and that euclidean spheres are not contractible.

Denote as direction D at p an oriented segment $T^+(p, p_1)$ with $pp_1 = \min\{\rho(p)/2, 1\}$. A topology for the directions is given by the metric $\varepsilon(D, D^1) = pp' + p_1 p_1'$, where $D' = T^+(p', p_1')$.

If a half geodesic $x(t)$ $(t \geqq 0, x(0) = p)$ begins with the direction D and is not a ray then it contains a maximal segment $T(p, m)$ and m is the so-called minimum point of p in the direction D. We put $\phi(D) = pm$, or, if $x(t)$ is a ray, $\phi(D) = \infty$. The union of all minimum points of p for variable D is the minimum (or cut) locus $M(p)$ of p. The distance $pM(p)$ is the function $\delta(p)$ of (12(e)).

It is very easily seen that $\phi(D)$ and $\delta(p)$ are upper semicontinuous functions of D and p respectively. In (sufficiently smooth) Riemann spaces they are continuous (see Bishop and Crittenden ⟨1, Section 11.6⟩); moreover, if m is a minimum point of p then p is one of m (in the obvious direction). This means in our terminology, if v with (pmv) does not exist then neither does s with (spm).

It is not known whether any of these three assertions is correct for G-spaces (the last holds trivially when $T(p, m)$ is not unique). If any of them should be false, it would provide another example of the type discussed at the end of the next section, namely theorems into whose formulation differentiability does not enter, but which fail to hold without it.

5. Differentiability

G-spaces include Finsler spaces in the standard sense (G, Section 15) but also many other spaces which are not. It is natural to ask for conditions on the distance xy which allow us to decide whether a given

G-space is a Finsler space. An answer to this question is important also from the classical point of view (see below).

The underlying idea is quite simple[9]. We express in a naive way that differentiability of a G-space at p must mean linearity of the distance up to terms of higher order. To formulate this in our context we denote, if $T(a, q)$ is unique, by $a_{q\beta}$ $(0 \leqq \beta \leqq 1)$ the point of $T(a, q)$ with $q a_{q\beta} = \beta q a$. In a straight Minkowski space we have $a_{q\beta} b_{q\beta}/\beta ab = 1$ for $a \neq b$. We require that this ratio be close to 1 when a, b, q are close together. Precisely we say:

The G-space R is differentiable at p if for a given $\varepsilon > 0$ a $\delta(\varepsilon) > 0$ $(\delta \leqq \rho(p))$ exists such that for a, b in $S(p, \delta)$

(1) $$|a_{p\beta} b_{p\beta} - \beta ab| \leqq \varepsilon \beta ab \quad \text{for } 0 \leqq \beta \leqq 1.$$

The space is continuously differentiable at p if $\delta_1(\varepsilon) > 0$ exists such that for a, b, q in $S(p, \delta_1)$

(2) $$|a_{q\beta} b_{q\beta} - \beta ab| \leqq \varepsilon \beta ab \quad \text{for } 0 \leqq \beta \leqq 1.$$

We will see that (2) is stronger, (1) may hold and (2) not. The terminology is motivated by the observation that for $f(t)$ $(|t| < \alpha)$ the existence of λ with $|f(s) - f(t) - \lambda(s-t)| \leqq \varepsilon |s-t|$ for $|s|, |t| < \delta(\varepsilon)$ implies $f'(0) = \lambda$ and that $f'(t)$ is continuous at 0 if it exists in a neighborhood of 0. Thus we are taking a liberty analogous to calling $f(t)$ continuously differentiable at 0 if the last inequality holds, although $f'(t)$ may not exist for all $t \neq 0$. We prove

(3) *Theorem. Let the G-space R be differentiable at p. Then*

(a) $$m(a, b) = \lim_{\beta \to 0} \frac{a_{p\beta} b_{p\beta}}{\beta} \quad (a, b \in S(p, \rho(p)))$$

exists and defines in $S(p, \rho(p))$ a metric equivalent to ab.

(b) *$m(a, b) = ab$ when a, b, p are collinear,*

(c) *$m(a_{p\beta}, b_{p\beta}) = \beta m(a, b)$,*

(d) *$|m(a, b) - ab| \leqq \varepsilon ab$ for $a, b \in S(p, \delta(\varepsilon))$.*

Because p is distinguished we put $a_{p\beta} = a_\beta$ and notice first that

(4) $$(a_{\beta_1})_{\beta_2} = a_{\beta_1 \beta_2}.$$

Applying (1) to a_β, b_β we obtain for $0 < \alpha < \beta < 1$ and $pa_\beta < \delta(\varepsilon)$ hence for $\beta < \delta(\varepsilon)/\rho(p)$ that

$$|a_\alpha b_\alpha - \beta^{-1} \alpha a_\beta b_\beta| \leqq \varepsilon \beta^{-1} \alpha a_\beta b_\beta.$$

---

[9] We follow Busemann ⟨5⟩. An earlier solution is found in Busemann [1], which was not discussed in G because it seemed (and proved) too involved.

Thus for $\beta<\beta_0<\delta(2^{-1})/\rho(p)$

(5) $$2^{-1}\beta_0^{-1}a_{\beta_0}b_{\beta_0}\leqq\beta^{-1}a_\beta b_\beta\leqq 3\cdot 2^{-1}\beta_0^{-1}a_{\beta_0}b_{\beta_0}$$

and for $\beta<\min(\delta(\varepsilon)/\rho(p), \beta_0)$

$$|\alpha^{-1}a_\alpha b_\alpha-\beta^{-1}a_\beta b_\beta|\leqq\varepsilon\beta^{-1}a_\beta b_\beta\leqq\varepsilon 3\cdot 2^{-1}\beta_0^{-1}a_{\beta_0}b_{\beta_0}.$$

This inequality proves the existence of $m(a,b)=\lim\beta^{-1}a_\beta b_\beta$ and also that $0<m(a,b)=m(b,a)<\infty$ for $a\neq b$. The triangle inequality is obvious and the equivalence of $m(a,b)$ and ab is obtained from (5) for $\beta\to 0^+$.

(b) is clear and (c) follows from (4). Dividing (1) by β and letting $\beta\to 0^+$ gives (d).

We extend the metric $m(a,b)$ which is defined in $S(p,\rho(p))$ by adding new points which are not in R. Let $\rho'=\rho(p)/2$ and $a\in K(p,\rho')$. We define a_β for $0\leqq\beta\leqq 1$ as before, for $\rho'<\beta<2$ as the point in $S(p,\rho(\beta))$ for which $(p a a_\beta)$ and $pa:pa_\beta=\rho':\beta$ and as a new point for $\beta\geqq 2$. The set R_p of all a_β is metrized by

$$m(a_\alpha, b_\beta)=\delta^{-1}m(a_{\alpha\delta}, b_{\beta\delta}),$$

where $\delta>0$ is chosen so small that $\alpha\delta$ and $\beta\delta$ are less than $\rho(p)$. Because of (c) this is independent of δ and leads to the previous $m(a_\alpha, b_\beta)$ when defined. Moreover, the relations

(6) $$(a_{\beta_1})_{\beta_2}=a_{\beta_1\beta_2} \quad\text{and}\quad m(a_{\delta\alpha}, b_{\delta\beta})=\delta m(a_\alpha, b_\beta)$$

hold for any nonnegative β_1, β_2, α, β, δ in the space R_p which we call the normal tangent space of R at p. We prove

(7) *R_p satisfies the Axioms II, III and IV with $\rho(p)=\infty$.*

To prove II or finite compactness let $m(a^\nu_{\alpha_\nu}, p)<M$ $(a^\nu_{\alpha_\nu}=(a^\nu)_{\alpha_\nu})$ and $0<\delta<2^{-1}M^{-1}\rho(p)$. Then $a^\nu_{\delta\alpha_\nu}\in S(p,\rho(p)/2)$ and $\{a^\nu_{\delta\alpha_\nu}\}$ has an accumulation point b in $\bar S(p,\rho(p)/2)$ (for the original metric xy and hence also for $m(x,y)$). We conclude from (6) that $b_{\delta^{-1}}$ is an accumulation point of $a^\nu_{\alpha_\nu}$.

We prove III and IV simultaneously by showing that for given distinct a, b in R_p and given $\rho>0$, $(\rho\neq 1)$ there is a point c with

$$m(a,c)=\rho\, m(a,b) \quad\text{and}\quad \begin{cases} m(a,c)+m(c,b)=m(a,b) & \text{if } \rho<1\\ m(a,b)+m(b,c)=m(a,c) & \text{if } \rho>1.\end{cases}$$

For a suitable $\delta_0>0$ and any $\delta<\delta_0$ there is a point d^δ with

$$a_\delta d^\delta=\rho\, a_\delta b_\delta \quad\text{and}\quad \begin{cases} a_\delta d^\delta+d^\delta b_\delta=a_\delta b_\delta & \text{if } \rho<1\\ a_\delta b_\delta+b_\delta d^\delta=a_\delta d^\delta & \text{if } \rho>1.\end{cases}$$

Define $c^\delta=(d^\delta)_{\delta^{-1}}$. Then

$$m(p,d^\delta)=pd^\delta\leqq pa_\delta+a_\delta d^\delta=pa_\delta+\rho\, a_\delta b_\delta$$
$$\leqq(1+\rho)\,pa_\delta+\rho\, pb_\delta=[(1+\rho)\,m(p,a)+\rho\, m(p,e)]\;\delta,$$

so that $m(p,c^\delta)=\delta^{-1}m(p,d^\delta)$ is bounded. We show, and emphasize for a later application, that every point c which is the limit of a converging sequence c^{δ_ν} with $\delta_\nu\to 0$ (i.e., $m(c,c^{\delta_\nu})\to 0$), satisfies the assertion.

For simplicity put $a_\nu=a_{\delta_\nu}$, $b_\nu=b_{\delta_\nu}$, $c_\nu=c_{\delta_\nu}$, $c^\nu=c^{\delta\nu}$, $d^\nu=d^{\delta\nu}$. For large ν

$$a_\nu d^\nu-c_\nu d^\nu\leqq a_\nu c_\nu<a_\nu d^\nu+d^\nu c_\nu$$

and by (3 d)

$$m(c,d^\nu)=\delta_\nu^{-1}\,m(c_\nu,d^\nu)\geqq 2^{-1}\,\delta_\nu^{-1}\,c_\nu d^\nu,$$

so

$$m(a,c)=\lim\delta_\nu^{-1}\,a_\nu c_\nu=\lim\delta_\nu^{-1}\,a_\nu d^\nu=\rho\,\lim\delta_\nu^{-1}\,a_\nu b_\nu=\rho\, m(a,b).$$

The relations $m(a,c)+m(c,b)=m(a,b)$ if $\rho<1$, or $m(a,b)+m(b,c)=m(a,c)$ if $\rho>1$, are proved by the same method.

If R is differentiable at p it is called *regular* at p if R_p satisfies Axiom V, i.e., uniqueness of prolongation.

When R is a Finsler space derived from an integrand $F(x,\xi)$ (see G, Section 15) and p has local coordinates x_0^i, this definition of regularity coincides with the usual one, which requires that $F(x_0,\xi)=1$ be a strictly convex hypersurface in ξ-space. It is always assumed in the theory of Finsler spaces that (compare condition (k) in G p. 84) the space is everywhere regular. This justifies to a certain extent introducing regularity here. But *there are integrands of class C^∞ which are nowhere regular and lead to G-spaces* (see Section 16), so that regularity is not a natural requirement in our context. It can be avoided. We require it nevertheless, because all known approaches without regularity are very long with uninteresting details.

We would like to prove that the metric $m(a,b)$ is Minkowskian. However, in Section 6 we will exhibit examples which demonstrate that R_p (even if regular) need not be Minkowskian. But:

(8) *Theorem. If the G-space R is continuously differentiable and regular at p, then the normal tangent space R_p is Minkowskian.*

It suffices to show that for any three points, a,b,c in R_p and the midpoints b' of a,b and c' of a,c the relation

$$m(b,c)=2m(b',c')$$

holds, see G p. 261. Because of (6) we may assume that a,b,c lie in $S(p,\rho(p))$.

Let d^β be the midpoint of a_β and b_β in the metric xy and f^β that of a_β and c_β. In the preceding proof we saw that with respect to $m(x, y)$ the points $(d^\beta)_{\beta^{-1}}$ $(0<\beta\leqq 1)$ form a bounded set and that any converging sequence $(d^{\beta_\nu})_{\beta_\nu^{-1}}$ $(\beta_\nu \to 0+)$ tends to the midpoint b' of a and b (for $m(x, y)$) and the same holds for f^{β_ν} so $\lim_{\beta\to 0+} (d^\beta)_{\beta^{-1}} = b'$, $\lim_{\beta\to 0+} (f^\beta)_{\beta^{-1}} = c'$.

From continuous differentiability we conclude $\lim 2d^\beta f^\beta / b_\beta c_\beta = 1$ for $\beta \to 0+$ or $m(b, c) = \lim \beta^{-1} b_\beta c_\beta = 2 \lim \beta^{-1} d^\beta f^\beta$. But

$$b'_\beta c'_\beta - d^\beta b'_\beta - f^\beta c'_\beta \leqq d^\beta f^\beta \leqq b'_\beta c'_\beta + d^\beta b'_\beta + f^\beta c'_\beta$$

and

$$\beta^{-1} d^\beta b'_\beta \leqq 3 (d^\beta)_{\beta^{-1}} b'$$

by (5). Therefore

$$\beta^{-1} d^\beta b'_\beta \to 0, \qquad \beta^{-1} f^\beta c'_\beta \to 0$$

and so

$$m(b, c) = 2 \lim \beta^{-1} d^\beta f^\beta = 2 \lim \beta^{-1} b'_\beta c'_\beta = 2m(b', c').$$

As a corollary we notice

(9) *If a G-space is continuously differentiable and regular at one point then it is a topological manifold.*

For, $m(x, y)$ and xy provide topologically equivalent metrizations of $S(p, \rho(p))$ which by (3b) is a sphere with the same radius for $m(x, y)$ and as such homeomorphic to E^n with some finite n. We know that any distinct points of a G-space have homeomorphic neighborhoods, see G (10.1).

Consider a Finsler space given as a manifold of a certain class C^k $(k \geqq 4)$ with an $F(x, \xi)$ of class C^{k-1} [10]. By passing to nonadmissible coordinates $u^1, \ldots, u^n$ we can destroy all smoothness properties of the metric in terms of u. The distance may no longer be expressible as an integral of a function $F^*(u, \eta)$ over a segment. The important question arises how we can *recognize from the intrinsic distance that this singular behavior is due to a poor choice of coordinates.* This problem is analogous to (but, of course, far easier than) Hilbert's Problem V on continuous groups and is quite inaccessible in the classical approach.

Our conditions for continuous differentiability and regularity do not use coordinates and will therefore be satisfied. Normal coordinates at p are – following the usual terminology – defined in $S(p, \rho(p))$ as affine coordinates with p as origin belonging to the normal Minkowski space, R_p, defined by Theorem (8). These coordinates will be of class

[10] We observe that the order $k \geqq 4$ in G Section 15 can be replaced by $k \geqq 3$ using the approach of Carathéodory ⟨1⟩ through partial differential equations. In the present discussion $k \geqq 4$ is preferable.

C^{n-2} except at the origin where they are in general only of class C^1 (see Busemann ⟨2⟩). In the Riemannian case they are everywhere of class C^{n-2}.

If q is any given point of R we can (because of $\lim_{p\to q}\rho(p)=\rho(q)>0$) find $p\neq q$ such that q lies in $S(p,\rho(p))$. Using normal coordinates with origin p we obtain local coordinates about q in terms of which the space is of class C^{n-2} and the induced $F(x,\xi)$ of class C^{n-3}. Thus we obtain a complete answer in the case of class C^∞ and recover at least C^{n-2} when the original space was of class C^n.

The problem may be put differently: *given a G-space which is everywhere continuously differentiable and regular to find out whether it is a Finsler space of a certain class.* We produce coordinates about q as before. If q lies in two different normal coordinate systems with origins p_1, p_2, $(p_i\neq q)$ then the transition functions from one system to the other will be of class C^n if the space is with suitable coordinates a manifold of class C^{n+2}, and its metric is obtained from a $F(x,\xi)$ of class C^{n+1}. Thus the method allows us to establish class C^n, when the space is optimally of class C^{n+2}. For class C^∞ we have again a complete solution.

These considerations show that continuous differentiability of G-spaces is significant for the transition from G-spaces to Finsler spaces and even for the classical theory. But does it properly belong to G-spaces, in other words, *are there questions which arise naturally in G-spaces which have a negative answer without, and an affirmative answer with, differentiability assumptions?*

Section 6 *will exhibit such a case:* A similarity with factor $k>0$ is a map ϕ of a metric space on itself with $\phi x\,\phi y=k\,x\,y$ for all x, y.

A similarity with factor $k\neq 1$ of a complete space has exactly one fixed point. (For this and the following see Section 6.)

One would expect that a G-space is Minkowskian, if similarities exist for all factors k. Actually the space may have infinitely many points each of which is the fixed point of similarities with given k without being Minkowskian. But it is a Minkowski space if it possesses a single similarity with factor $k\neq 1$, and is continuously differentiable at its fixed point.

There are other examples which seem simpler but are less convincing. In a Finsler space small spheres are convex, i.e., for a given point p there is a positive $\eta\leqq\rho(p)$ such that $T(q,r)$ lies in $S(p,\eta)$ with q, r, see G, p. 162.

This is not necessarily correct even when the affine lines are the geodesics. Let $\phi(t)=0$ for $t\leqq 0$, $=t^{\frac{1}{2}}$ for $0<t<1$, and $=1$ for $t\geqq 1$. One easily checks that with the distance

$$x\,y=[(x_1-y_1)^2+(x_2-y_2)^2]^{\frac{1}{2}}+|\phi(x_1)-\phi(y_1)|+|\phi(x_2)-\phi(y_2)|$$

the (x_1, x_2)-plane becomes a G-space with the ordinary lines as geodesics and that no circle about (0, 0) with radius less than 2 is convex.

However, the proofs for the existence of η rest in the last analysis on the fact that the normal curvatures of the local unit spheres are bounded away from 0 and ∞. Restrictions on the upper and lower one-sided normal curvatures probably suffice to draw the same conclusion, so that second derivatives are most likely avoidable and quite conceivably a hypothesis involving no derivatives, but always satisfied under the classical assumptions will suffice.

The problems mentioned at the end of the last section may be of a similar nature if their answers are negative for general G-spaces.

II. Desarguesian Spaces

An n-dimensional desarguesian space is a metrization of an open subset of the n-dimensional projective space P^n as a G-space R whose geodesics fall on projective lines. R is either the entire P^n and the geodesics are all isometric to one circle, or R is straight and may be regarded as an open convex set C of the n-dimensional affine space A^n with the intersections of the affine lines with C as geodesics, see Sections 12–14 in G or (8.2) here.

The present chapter contains new contributions to this topic. First we construct spaces, desarguesian and nondesarguesian which show that *in* (5.8) *differentiability cannot be substituted for continuous differentiability.* These spaces possess similarities with given factors k, in one of them all points of a hyperplane are centers of similarities. Owing to the importance for the axiomatics of G-spaces explained above we investigate in detail, what can be deduced from the existence of similarities alone and where continuous differentiability enters to establish the Minkowskian character of the metric.

Next we show that an *n-dimensional desarguesian space can be imbedded as a hyperplane in an $(n+1)$-dimensional desarguesian space.* The problem was left open in G for $R=P^n$. We also report on an analogous problem for area.

We conclude the chapter by *characterizing Hilbert's and Minkowski's geometries* among all straight desarguesian spaces by the simple and appealing property that *an isometry of one line on another is a projectivity.* Under differentiability hypotheses this is a special case of a theorem of Berwald ⟨1⟩ on desarguesian spaces with constant Finsler curvature.

6. Similarities

To construct examples in connection with (5.8) consider the affine (x^1, x^2)-plane. A line can be written uniquely as

$$x^1 \cos\alpha + x^2 \sin\alpha - p = 0, \qquad p \geqq 0, \ -\pi/2 < \alpha \leqq \pi/2.$$

Let $g(t)$ be a continuous strictly increasing function on $[0,\infty)$ with $g(0)=0$ and $g(t)\to\infty$ for $t\to\infty$. Put

$$f(x,\alpha) = \operatorname{sign}(x^1\cos\alpha + x^2\sin\alpha)\, g(|x^1\cos\alpha + x^2\sin\alpha|), \quad -\pi/2 < \alpha \leqq \pi/2.$$

We claim that

$$(1)\qquad \rho(x,y)=\int_{(-\pi/2,\pi/2]} |f(x,\alpha)-f(y,\alpha)|\,d\alpha$$

defines a distance for which the plane becomes a straight G-space with the affine lines as geodesics. It possesses all rotations about 0.

It is obvious that $0<\rho(x,y)=\rho(y,x)<\infty$ for $x\neq y$. The triangle inequality follows from

$$(2)\qquad |f(x,\alpha)-f(z,\alpha)|\leqq|f(x,\alpha)-f(y,\alpha)|+|f(y,\alpha)-f(z,\alpha)|.$$

If $y=(1-t)\,x+t\,y$ $(0<t<1)$ then for each α either

$$f(x,\alpha)\geqq f(y,\alpha)\geqq f(z,\alpha) \quad\text{or}\quad f(x,\alpha)\leqq f(y,\alpha)\leqq f(z,\alpha).$$

This is seen by interpreting x^1, x^2 as cartesian coordinates of a euclidean plane and remembering that $x^1\cos\alpha+x^2\sin\alpha$ is the signed distance of the line with normal α through x from $0_p=(0,0)$. Therefore $f(x,\alpha)$ changes monotonically with the line normal to α on which x lies, whence

$$\rho(x,y)+\rho(y,z)=\rho(x,z).$$

If y does not lie on the affine segment from x to z then α_0 exists such that either

$$f(y,\alpha_0)>\max\{f(x,\alpha_0),f(z,\alpha_0)\} \quad\text{or}\quad f(y,\alpha_0)<\min\{f(x,\alpha_\alpha),f(z,\alpha_0)\}$$

and this is true in a neighborhood of α_0. Thus inequality holds in (2) for an α-interval, whence $\rho(x,y)+\rho(y,z)>\rho(x,z)$.

This shows that the affine segments and only these are metric segments.

The euclidean rotations about 0 clearly are rotations for the metric $\rho(x,y)$. To prove finite compactness, it therefore suffices to show that $\rho(0_p,(t,0))\to\infty$ for $t\to\infty$, which follows from

$$\rho(0_p,(t,0))>\int_0^{\pi/4} g(t\cos\alpha)\,d\alpha\geqq\pi\,g(2^{-\frac{1}{2}}t)/4.$$

We apply this result to the function $g(t)=t^\beta$ $(\beta>0)$ and denote the corresponding distance by $\rho_\beta(x,y)$. Since $f(h\,x,\alpha)=h^\beta f(x,\alpha)$ for $h>0$ we see that for a given $k>0$

$$(3)\qquad \rho_\beta(\phi\,x,\phi\,y)=k\,\rho_\beta(x,y) \quad\text{if } \phi\,x=k^{1/\beta}\,x.$$

ϕ is a similarity with factor k and center 0_p. Also, the plane with the metric ρ_β is clearly differentiable at p, it is, in fact, its own normal tangent space at p and therefore is also regular at p. The metric is Minkowskian only for $\beta=1$ in which case it is euclidean. It is easy to verify directly that (5.2) is not satisfied for $\beta\neq1$.

Before continuing with (5.8) we point out that other very interesting phenomena can be produced by suitable choices of $g(t)$.

(4) *There are metrizations $\rho(x, y)$ of the affine plane as a G-space with the affine lines as geodesics such that all rotations about one point z exist and*

$$\lim \rho(x, L)=0 \quad \textit{for } x\in L', \ \rho(z, x)\to\infty \ \textit{whenever } L\|L'$$ [1]

or also

$$\lim \rho(x, L)=\infty \quad \textit{for } x\in L', \ \rho(z, x)\to\infty, \textit{ whenever } L\|L'.$$

The first case enters for $g(t)=\log(1+t)$, the second for $g(t)=e^t-1$. The estimates are simple in the latter case and can be found in Busemann ⟨6⟩ for the former.

To obtain *nondesarguesian* examples for similarities and differentiability we consider in the (x^1, x^2)-plane the well known system W of curves defined by Moulton. W consists of

1) the lines $x^2=(x^1-a)\tan\alpha$, $\pi/2<\alpha\leqq\pi$
2) $x^2=\text{const}$ or $\alpha=\pi/2$
3) the broken lines

$$x^2=\begin{cases}2(x^1-a)\tan\alpha & \text{for } x^1\leqq a\\ (x^1-a)\tan\alpha & \text{for } x^1>a\end{cases}, \quad 0<\alpha<\pi/2.$$

In all cases we call α the inclination of the curve. W satisfies the parallel axiom (two curves are parallel if and only if they have the same inclination), but not the Theorem of Desargues, see for example Hilbert ⟨1⟩.

We introduce the euclidean distance $e(x, y)=[(x^1-y^1)^2+(x^2-y^2)^2]^{\frac{1}{2}}$ and denote for two points x, y by $\delta_\alpha(x, y)$ the euclidean distance of the lines in W with inclination α through x resp. y. Then

$$\delta_\alpha(x, y)\leqq e(x, y) \quad \text{and} \quad \delta_\alpha(kx, ky)=k\,\delta_\alpha(x, y) \quad \text{for } k>0.$$

Therefore

$$\delta(x, y)=\int_{(0,\pi]}\delta_\alpha(x, y)\,d\alpha$$

is finite and satisfies

$$\delta(kx, ky)=k\,\delta(x, y) \quad \text{for } k>0. \tag{5}$$

Moreover, δ_α and hence δ is invariant under translations parallel to the x^1-axis. The argument that δ defines a G-space with the curves in W as geodesics is very similar to the proof of (1) [2].

It follows from (5) that ϕ: $x\to kx$ $(k>0)$ is a similarity with factor k and center 0_p and from translation invariance that every *point of the*

[1] This shows that the answer to Problem (19), G p. 404 is negative.

[2] Examples (3) and (5) show that the assertion (22) in G p. 405 is not correct.

x^1-axis is the center of similarities with arbitrary k. If $\bar{x}=(x^1, x^2, x^3, \ldots, x^n)$ then

$$\bar{x}\,\bar{y}=\left[\delta^2(x, y)+\sum_{i=3}^{n}(x^i-y^i)^2\right]^{\frac{1}{2}}$$

extends $\delta(x, y)$ to a metrization of the $\bar{x}$-space as a straight space (see G (8.15)) which is not desarguesian and where each point in the hyperplane $x^2=0$ is the center of similarities with arbitrary $k>0$.

The space is differentiable at the origin and is again its own normal tangent space, hence regular, but is, of course, not continuously differentiable.

Besides the global similarities which we have hitherto *considered* we define a *local similarity* with factor $k>0$ of a metric space R as a map ϕ of R onto itself with the property that each point p has a neighborhood $S(\rho, \beta)$ which ϕ maps on $S(\phi p, k\beta)$ such that $\phi x\,\phi y=k\,x\,y$ for x, y in $S(p, \beta)$; here k is assumed to be independent of p, but if R is a G-space this can be proved.

(6) *If ϕ is a local similarity with factor k of a space with an intrinsic metric then $\phi x\,\phi y\leqq k\,x\,y$.*

Obviously $\lambda(\phi C)=k\,\lambda(C)$ for any rectifiable curve. For given points x, y let C be a curve from x to y with $\lambda(C)<x\,y+\varepsilon$. Then

$$\phi x\,\phi y\leqq\lambda(\phi C)=k\,\lambda(C)\leqq k(x\,y+\varepsilon).$$

The inverse of a similarity ϕ with factor k is defined and has factor k^{-1}. The inverse of a local similarity is in general not defined. The product of two (local) similarities with factors k_1, k_2, is a (local) similarity with factor $k_1 k_2$. We remember the following lemma from functional analysis:

(7) *If R is a complete metric space and ϕ maps R into itself with $\phi x\,\phi x\leqq k\,x\,y$ $(k<1)$, then ϕ has exactly one fixed point f and $\lim_{\nu\to\infty}\phi^\nu x=f$ for all x.*

For, putting $u_\nu=\phi^\nu u$ we have

$$x_\nu y_\nu\leqq k^\nu x\,y, \qquad \text{in particular } x_\nu x_{\nu+1}\leqq k^\nu x_0\,x_1$$

so that $\{x_\nu\}$ is a Cauchy sequence for any x tending to some point f and y_ν also tends to f. An immediate consequence of these remarks is:

(8) A similarity with factor $k\neq 1$ of a complete space has exactly one fixed point f and for every point x and $\nu\to\infty$

$$\phi^\nu x\to f \quad \text{if } k<1 \quad \text{and} \quad \phi^{-\nu}x\to f \quad \text{if } k>1.$$

If the space is complete and the metric is intrinsic then a local similarity ϕ with factor $k<1$ has exactly one fixed point f and $\phi^\nu x\to f$ for each x.

(9) *A compact space with an intrinsic metric (and more than one point) does not possess local similarities with factor $k<1$.*

This follows from (6). Observe that local similarities with factor $k>1$ may exist. The most trivial example, which also shows that several fixed points are possible, is the map $e^{i\phi} \to e^{3i\phi}$ of the unit circle on itself.

Let R_k denote the space obtained from R by multiplying all distances by k. Then a local similarity ϕ of R with factor k may be interpreted as a local isometry of R_k on R. Therefore we deduce from the theory of covering spaces, see G (27.17):

(10) *If R is a G-space whose fundamental group is not isomorphic to a proper subgroup then a local similarity of R is global.*

We now prove

(11) *If a G-space R possesses a local similarity ϕ with factor $k<1$, then R is straight and ϕ is global.*

Since straight spaces are simply connected it suffices by (10) to prove that R is straight. This will be the case if the geodesic curve connecting two given points x, y is unique. For then the geodesic through distinct points is unique and is by G (9.9) a straight line.

Let $z(t)$ $(\alpha \leqq t \leqq \beta)$ be any geodesic curve from x to y. Choose ν so large that

$$z_\nu(t) = \phi^\nu z(t) \in S = S(f, \rho(f)/3) \quad \text{for } t \in [\alpha, \beta],$$

where f is the fixed point of ϕ. Then $z_\nu(t)$ is a geodesic curve from x_ν to y_ν in S because locally

$$z_\nu(t_1)\, z_\nu(t_2) = \nu^{-1}(t_2 - t_1) \quad \text{for } t_1 < t_2.$$

But the segment $T(x_\nu, y_\nu)$ is the only geodesic curve from x_ν to y_ν in S. (Every half geodesic with origin x_ν begins with a segment $T(x_\nu, u)$ of length $\rho(x_\nu) > 2\rho(f)/3$ and $uf > \rho(f)/3$ hence $u \notin S$.)

For $k>1$ we have:

(12) *If the G-space R admits a local similarity ϕ with factor $k>1$, then its universal covering space $\bar{R}$ is straight and possesses a similarity $\bar{\phi}$ which lies over ϕ and has the same factor k.*

ϕ has fixed points and if f is a fixed point of ϕ then $\bar{\phi}$ can be chosen such that its (unique) fixed point lies over f.

As in G, Chapter IV, we assume that $\bar{R}$ is related to R by a definite local isometry Ω and realize the fundamental group of R as the group of motions of $\bar{R}$ which lie over the identity of R. The existence of $\bar{\phi}$ over ϕ means that $\Omega\bar{\phi} = \phi\,\Omega$ and follows exactly as in the proof of G (25.7) or also from the general Covering Homotopy Theorem. Clearly $\bar{\phi}$ is a local similarity of $\bar{R}$ with factor k. Because of (10) $\bar{\phi}$ is global and by (11) applied to $\bar{\phi}^{-1}$ the space $\bar{R}$ is straight.

By (8) the fixed point $\bar{f}$ of $\bar{\phi}$ is unique and $\phi\Omega\bar{f}=\Omega\bar{\phi}\bar{f}=\Omega\bar{f}$, so $\bar{\phi}$ leaves $\Omega\bar{f}$ fixed. The last statement is easily proved by choosing f as the distinguished point in the proof of G (28.7).

These theorems lead us as far as we can go without continuous differentiability. With this concept we obtain the following principal result on similarities:

(13) *Theorem. Let the G-space R possess a local similarity with factor $k \neq 1$ and let R be continuously differentiable at a fixed point of ϕ. When $k<1$ then R is Minkowskian, when $k>1$ then the universal covering space of R is Minkowskian.*

(11) and (12) reduce (13) to the following:

(14) *If the straight space R possesses a similarity ϕ with factor $k \neq 1$ and is continuously differentiable at the fixed point f of ϕ, then R is Minkowskian.*

We may assume that $k<1$.

As in the proof of (5.8) we show that any three points a, b, c and the midpoints b' of a, b and c' of a, c satisfy $2b'c'=bc$.

We assume $b \neq c$ and have

$$b'_\nu c'_\nu/b_\nu c_\nu = k\, b'_{\nu-1} c'_{\nu-1}/k\, b_{\nu-1} c_{\nu-1} = b'_{\nu-1} c'_{\nu-1}/b_{\nu-1} c_{\nu-1}.$$

The points a_ν, b_ν, c_ν, b'_ν, c'_ν tend to f. Now b'_ν is the midpoint of a_ν, b_ν and c'_ν that of a_ν, c_ν. Therefore the definition of continuous differentiability (5.2) at f gives with $\beta=\frac{1}{2}$

$$2b'c'/bc = \lim 2b'_\nu c'_\nu/b_\nu c_\nu = 1.$$

7. Imbeddings of Desarguesian Spaces

It is proved in G p. 81 that an n-dimensional straight desarguesian space can be imbedded as a hyperplane in an $(n+1)$-dimensional desarguesian space, but the corresponding problem for $R=P^n$ is left open (G p. 403, Problem (10)). We will settle this question now and thus obtain the

(1) *Theorem. For a given n-dimensional desarguesian G-space R there is an $(n+1)$-dimensional desarguesian space R^* such that R is a hyperplane in R^* and the restriction of the metric of R^* to R is the given metric in R.*

The method of proof follows Busemann ⟨4⟩ where it is also used to give a simple proof for another problem ((9) G p. 403) left open in G which was first solved by Skornyakov ⟨1⟩. A still simpler proof for the latter problem will be given in Section 14.

Let R be a metrization of P^n with the projective lines as geodesics. These all have the same length α (G (13.1), (14.1) or (31.2)). We may assume $n \geqq 2$, because the assertion is trivial for $n=1$.

Then the n-dimensional sphere S is the universal covering space of R and the metric of R induces on S a spherelike metric (see G p. 129), i.e., the great circles of S are also great circles in the sense of the induced distance xy (G p. 25), all have length 2α, each semi-great circle is a segment of length α.

We consider S as the equator of an $(n+1)$-dimensional sphere S^*. In the proof integrals over subsets of S^* will occur, they all refer to the ordinary spherical measure in S^*.

We want to metrize S^* as a spherelike space with the great circles as geodesics, such that distances in S are preserved. Let H be one of the open hemispheres of S^* bounded by S and take a function $f(p)$ defined and continuous on $S \cup H = \bar{H}$, positive on H, zero on S and with $\int f(p) = 1$.

A *semi-circle* K_p is an arc of a great circle in S^* leading from p to its antipode p'. For $x \neq p, p'$ there is precisely one K_p through x, which we denote by $K_p(x)$. If $p \in H$ and $x \neq p, p'$, put

$$x_p = K_p(x) \cap S, \qquad \text{so that } x_p = x \text{ for } x \in S.$$

Observe that $x \to x_p$ maps antipodes in S^* on antipodes in S and a great circle in S^* on one in S.

We define

$$\text{(2)} \qquad \delta_p(x, y) = x_p y_p \quad \text{for } x \neq p, p', \; y \neq p, p'.$$
$$\delta_p(p, x) = \delta_p(p', x) = \delta_p(x, p) = \delta_p(x, p') = 0.$$

Then $\delta_p(x, y) \leqq \alpha$. Therefore

$$\text{(3)} \qquad \delta(x, y) = \int_H \delta_p(x, y)\, f(p)$$

exists and is finite. We show that $\delta(x, y)$ solves the problem. Notice that there are many solutions depending on the choice of $f(p)$.

Obviously

$$\delta(x, y) = xy \quad \text{for } x, y \text{ in } S$$

and

$$\delta(x, x) = 0, \quad \delta(x, y) = \delta(y, x) > 0 \quad \text{for } x \neq y.$$

If x, y, z are distinct from p and p' then

$$\text{(4)} \qquad x_p y_p + y_p z_p \geqq x_p z_p \quad \text{or} \quad \delta_p(x, y) + \delta_p(y, z) \geqq \delta_p(x, z).$$

If M is the union of x, y, z and their antipodes then integration over $H-M$ yields

$$\delta(x, y)+\delta(y, z) \geqq \delta(x, z).$$

Thus $\delta(x, y)$ satisfies the axioms for a metric space. We show next that this metric induces the correct topology on S^* or that given $\varepsilon>0$ and q a neighborhood $U(q)$ exists such that

$$\delta(x, y)<\varepsilon \quad \text{for } x, y \text{ in } U(q).$$

Because $f(p)$ vanishes on S there is a neighborhood V of S with

$$\int_{V \cap H} f(p)<\varepsilon/3\alpha.$$

We choose $U(q)$ so that $\int_{U(q)} f(p)<\varepsilon/3\alpha$ and $\delta_p(x, y)<\varepsilon/3$ for x, y in $U(q)$ and $p \in H-V-U(q)$. Then any x, y in $U(q)$ satisfy

$$\delta(x, y) \leqq \int_{V \cap H} \alpha f(p)+\int_{U(q)} \alpha f(p)+\int_{H-V-U(q)} \delta_p(x, y) f(p)<\varepsilon.$$

A semicircle K with end points x, x' (x' the antipode to x) p, p' not on K, is, with the metric $\delta_p(x, y)$ a segment of length α. Hence, if G is the great circle containing K then

$$\delta(x, x')=\int_{H-G} \delta_p(x, x') f(p)=\alpha.$$

Moreover, if x, y, z lie in this order on K then x_p, y_p, z_p lie on the projection of K from p on S which is a semi great circle in S, so that $x_p y_p+y_p z_p= x_p z_p$ and integration over $H-G$ gives

$$\delta(x, y)+\delta(y, z)=\delta(x, z).$$

It follows that G is also a great circle in S^* with respect to the metric $\delta(x, y)$ and that points with distance α on G are also antipodes for S^*.

Finally, if x, y, z do not lie on a great circle then inequality holds in (4) for all p not on the great S^2 through x, y, z so that $\delta(x, y)+\delta(y, z)>\delta(x, z)$. This proves that S^* is spherelike with the ordinary great circles as geodesics.

Identifying antipodal points in S yields an $(n+1)$-dimensional projective space R^* containing R with the projective lines as geodesics and the original distances in R.

It may be asked *why we used S and S^* instead of R and R^**. The projection x_p $(x \neq p)$ of $x \in R^*$ on R is the intersection of the line through p and x with R. If we define $\delta_p(x, y)$ as in (2) and $\delta(x, y)$ as in (3), where $f(p)>0$ on R^*-R, $f(p)=0$ on R and $\int_{R^*-R} f(p)=1$, then $\delta(x, y)=xy$ on R and

$\delta(x, y)$ metrizes R^*, also $\delta(x, y)+\delta(y, z)>\delta(x, z)$ for noncollinear x, y, z, but not every projective line is a geodesic.

The study of desarguesian spaces with symmetric or nonsymmetric distances is the topic of Problem IV among the famous problems of Hilbert [2]. This suggests an investigation of those *r-dimensional areas in A^n for which the r-flats minimize area.* Little has been done in this direction, in particular in the large.

It is proved in G (18.14) that any nonempty open set D in A^n can be metrized as a G-space such that the intersections of the affine lines with D become the geodesics.

The analogous problem has been solved for r-areas and also an extension corresponding to Theorem (1) has been proved by Busemann ⟨8⟩. We will report the results here without proofs. *The interest lies partly in the completeness concepts.*

We consider A^n as an open hemisphere H in S^n with intersections of the great circles with H as lines. Then every line has two end points on the boundary C of H and any convex set K in A^n has a compact boundary B_K in $\bar{H}$.

Let D be a nonempty open convex set in A^n. An r-flat in D is the nonempty intersection of an r-flat in A^n with D. An *r-area* $(1 \leqq r \leqq n)$ in D is a function $\alpha(M) \geqq 0$ defined on all Borel sets M lying in r-flats of D, which is completely additive in each r-flat, vanishes on sets lying in $(r-1)$-flats, is positive and finite for a nondegenerate r-simplex W with vertices in D and depends continuously on the vertices of W. The α-area of an r-dimensional polyhedral surface P with r-faces $F_1, \ldots, F_k$ is defined as

$$\alpha(P)=\sum_{i=1}^{k} \alpha(F_i).$$

The properties of α guarantee that $\alpha(P)$ is invariant under subdivision of the faces of P.

The *r-flats minimize α-area* if for every closed oriented r-dimensional polyhedron with faces $F_0, \ldots, F_k$

$$\alpha(F_0) \leqq \sum_{i>0} \alpha(F_i) \quad \text{with inequality when } F_0 \neq \bigcup_{i>0} F_i.$$

An euclidean area induced by a euclidean metrization of A^n (invariant under translation) has this property, but is not interesting unless $D=A^n$, because it lacks completeness. An r-area α is *complete*, if $\alpha(K)=\infty$ for an r-dimensional convex set K in D for which $B_K \cap B_D$ contains a nonempty open subset of B_K. To obtain more significant results different types of completeness must be distinguished when $r>1$. We say that α is *strongly complete* (or complete in the euclidean sense) when $\alpha(K)=\infty$

for any r-dimensional convex set K with $B_K \cap B_D \neq \phi$, and that α is *hyperbolically complete*, when it is complete, but $\alpha(W)$ is finite for an r-simplex W of which r vertices lie in D and one on B_D, with an obvious interpretation if the latter falls on C[3]. Then the following holds:

(5) *Theorem. If D is any open convex subset of A^n, and $1 \leqq r \leqq n$ then D possesses a strongly complete r-area for which the r-flats minimize area. For $r > 1$ there exists a hyperbolically complete r-area with the minimizing property when $B_D \cap A^n$ is strictly convex or $D = A^n$*[4].

Just as for $r=1$ these areas are far from unique for $r>1$. The proof for $1<r<n$ rests on an idea of integral geometry: One defines a density for the $(n-r)$-flats in D and $\alpha(M)$ is the measure of the $(n-r)$-flats intersecting M (where M is a Borel set in an r-flat). This idea is of considerable value also in G-spaces, see Sections 14, 15.

The great variety of r-areas, which detracts perhaps from (5), makes the following imbedding theorem analogous to (1) all the more interesting:

(6) *Theorem. Let a complete r-area $(1 \leqq r \leqq n)$ be given in $D \subset A^n$ for which the r-flats minimize area. Then a convex set $D^+ \subset A^{n+1}$ with $D^+ \cap A^n = D$ and a complete r-area α^+ in D^+ exists for which the r-flats minimize area and $\alpha^+(M) = \alpha(M)$ for Borel sets in r-flats of D. For $r>1$ the area α^+ can be chosen hyperbolically complete if α has this property.*

The question whether α^+ can be made strongly complete with α $(r>1)$ is open except for $D = A^n$, where the answer is affirmative.

8. A Characterization of Hilbert's and Minkowski's Geometries

The great arbitrariness entering the choice of a desarguesian metric for a given convex set in A^n on the one hand and the naturalness of Minkowski's and Hilbert's geometries (see G, Section 18) on the other suggest the question, whether the latter metrics can be jointly distinguished by simple geometric property.

For this question it is important to also consider *desarguesian spaces with nonsymmetric distances*, in fact, most contributors to Hilbert's Problem IV have treated this general case. We define these spaces as follows (see Sections 1 and 2).

A not necessarily symmetric intrinsic distance with compact $\bar{S}^+(p, \rho)$ is defined in a nonempty open set D of the n-dimensional projective space P^n and $xy + yz > xz$ if x, y, z do not lie on a projective line.

[3] For $r=1$ completeness always means strong completeness and coincides with the usual concept when the α-length of a segment in D is taken as the distance of its endpoints.

[4] The restriction to strictly convex $B_D \cap A^n$ is probably unessential.

This definition implies:

$T(x, y)$ exists for given $x \neq y$ and hence must lie on the projective line $L(x, y)$ through x and y. For any projective line L either $L \cap D = \emptyset$ or $L \subset D$ or $L \cap D$ is a connected open subset of L. Therefore the maximal partial geodesic $x(t)$ $(t \in M)$ containing $T(x, y)$ is unique and is the set $L(x, y) \cap D$. Since the $\bar{S}^+(p, \rho)$ are compact $x(t)$ either traverses all of $L(x, y)$, hence is a periodic geodesic; or has the form

$$(1) \qquad x(t), \quad \alpha < t < \infty, \quad x(t_1)\, x(t_2) = t_2 - t_1 \quad \text{for } t_1 < t_2,$$

where α may or may not equal $-\infty$. Funk's geometry at the end of Section 1 provides an example when α is always finite.

The following important fact is due to Hamel [1]:

(2) *Theorem. Let the n-dimensional desarguesian space be defined in* $D \subset P^n$ $(n > 1)$. *Then either* $D = P^n$, *the projective lines are the geodesics and all have the same length traversed in either direction. Or D may be regarded as a convex subset of* A^n *and all maximal partial geodesics have the form* (1).

The proof is substantially the same as that for G-spaces in G p. 74, but we give it here because certain changes are necessary.

Let Ω be map of S^n on P^n which maps antipodes in S^n on the same point of P^n. Then $D' = \Omega^{-1} D$ is either connected or not.

In the first case it is intuitively clear and proved as in G p. 74 that D' is a covering space of D which has an intrinsic metric and compact $\bar{S}^+(p, \rho)$. Each maximal partial geodesic G in D' lies on a great circle C of S^n. If $p' \in C - G$ then its antipode $p'' \in C - G$. On the other hand if $q' \in G$ then the antipode q'' of q' lies in D' and hence on C and so does a point r close to q'' on C. But the only possible segment $T(q', r)$ is a subarc of C, which does not lie in D' since it contains p' or p''. Thus $D' = S^n$.

If q' and q'' are antipodes then each semi-great circle T from q' to q'' is a segment $T(q', q'')$ for D'. If not there is a largest subarc T_0 beginning at q' and ending at $r \neq q''$ say, which is a segment. The other arc A of C from q' to r must also be a segment because each subarc of A from q' to a point s close to r is a $T(q', s)$. Thus the subarc of A from q' to q'' is a segment. If C' is another great circle through q' and q'' then for the same reason one of the semi-great circles from q' to q'' of C' is a segment, $T(q', q'')$ but then $(q'\, q''\, r)$ contradicts (1.5).

Moreover, each of these semi-great circles is also a $T(q'', q')$ and of the same length, i.e., $q'\, q'' = q''\, q'$, because Ω maps a semi-great circle of C from q' to q'' and its continuation from q'' to q' on the same geodesic in D.

If D' is not connected it consists of two disjoint sets, D'_1, D'_2, with the property that if a point lies in D'_1 its antipode lies in D'_2. Thus D'_1 contains with any two points p, q $(p \neq q)$ the unique arc of the great circle through p and q which does not contain antipodes.

Consider S^n as a sphere with center c in E^{n+1} and form the union V of all euclidean rays with origin c and through points of D_1'. Then the intersection of V with any plane (= two-flat) through c is convex, so that V is convex. c is not an interior point of V (otherwise $D_1' = S^n$), hence V possesses at c a supporting hyperplane H^* and D_1' lies in one of the hemispheres of S^n bounded by $H^* \cap S^n$. Because D_1' is open in S^n it lies in the open hemisphere, so that $\Omega(H^* \cap S^n)$ is a hyperplane in P^n disjoint from D. That all maximal partial geodesics in D have the form (1) is obvious.

A joint characterization of Hilbert's and Minkowski's geometries is, under differentiability hypotheses, contained in the following two facts.

The (not necessarily symmetric) Minkowskian geometries and the (always symmetric) Hilbert geometries are the only noncompact desarguesian spaces with compact $\bar{S}^{\pm}(p, \rho)$ and constant Finsler curvature (Berwald ⟨1⟩ and Funk ⟨1⟩).

The desarguesian spaces with constant Finsler curvature are characterized by the property (Berwald ⟨1⟩, a simpler proof in Funk ⟨2⟩):

An isometry of a maximal partial geodesic on another or itself is a projectivity.

This means precisely: if $x(t)$ $(t \in M)$ and $y(t)$ $(t \in N)$ represent maximal partial geodescis lying on the projective lines L_x and L_y respectively and if a map ϕ of M on N exists such that

$$x(t_1)\,x(t_2) = y(\phi\, t_1)\, y(\phi\, t_2)\,^{5} \qquad \text{for } t_1 < t_2,\ t_i \in M,$$

then there is a projectivity ψ of L_x on L_y with

$$\psi\, x(t) = y(\phi\, t) \qquad \text{for } t \in M.$$

If the balls $\bar{S}^{\pm}(p, \rho)$ are compact then, under the present conditions, all maximal partial geodesics are geodesics. Therefore the quoted results yield with smoothness assumptions:

(3) *Theorem. Among all noncompact not necessarily symmetric desarguesian spaces with compact $\bar{S}^{\pm}(p, \rho)$ the geometries of Hilbert and Minkowski are distinguished by the property that an isometry of one geodesic on another or itself is a projectivity.*

We give here a simple direct proof without differentiability hypotheses. That the two geometries have the property is obvious, so we only need to prove sufficiency.

Because of Theorem (2) we may regard the space as an open convex subset D of A^n. Since an entire affine line cannot be mapped projectively

[5] This implies that ϕ is a topological map.

on a proper subset, either all geodesics are affine lines and $D = A^n$ or all are proper subsets of affine lines which have two distinct endpoints one of which may be at infinity.

Consider the second case first. Then D does not contain an entire affine line. Therefore, even if $\bar{D}$ is not compact there is a supporting hyperplane H of $\bar{D}$ which intersects $\bar{D}$ in a compact set. With respect to a hyperplane H' parallel to H and not intersecting $\bar{D}$ as ideal locus, the set $\bar{D}$ will be compact. So we may assume that to be the case.

A geodesic G has two distinct endpoints u, v on the boundary of D. On G we introduce the hyperbolic or Hilbert metric

$$h(a,b) = |\log R(a,b,u,v)|$$

where $R(\ )$ mean the cross ratio. A translation of L with respect to the given metric (i.e., a map $x(t) \to x(t+\alpha)$ if $x(t)$ represents G) induces by hypothesis a projectivity of the projective line L carrying G with fixed points u and v and is therefore also a translation for the metric $h(a,b)$. This implies

$$ab = k_G\, h(a,b)$$

where k_G is some positive factor of which we want to prove that it does not depend on G.

If G, G' are distinct geodesics on the lines L, L' and with end points u, v respectively u', v' then an isometry of G on G' induces a projectivity of L on L' which, with proper notation, maps u on u' and v on v'. Any two points a, b on G and their images a', b' on G' satisfy $ab = a'b'$ and $h(a,b) = h(a',b')$, therefore $k_G = k_{G'}$. Thus ab is a Hilbert metric in D.

The proof of G (18.5) shows that $xy = yz > xz$ for any noncollinear x, y, z if and only if the boundary of D does not contain proper coplanar segments whose convex hull intersects D[6].

We turn to the case $D = A^n$ and consider again a translation $x(t) \to x(t+\alpha)$ of a geodesic G in D.

The projective line L containing G originates from D by adding one point u and the translation induces a projectivity ϕ of L which has u as only fixed point. For any $q \in G$ the points $q, \phi^2 q, \phi q, u$ form a harmonic quadruple. Therefore ϕq is the affine midpoint of q and $\phi^2 q$. On the other hand, ϕq is also the midpoint of q and $\phi^2 q$ in the sense of the given metric ab. Thus affine midpoints and midpoints for ab coincide, which characterizes Minkowskian geometry. In fact, this is taken as the definition in G Section 17. The symmetry of the distance which is assumed there is not used to prove the invariance of the metric under the translations of A^n.

It should be emphasized that *in Theorem* (3) *the hypotheses that the* $\bar{S}^{\pm}(p,\rho)$ (*and not only the* $\bar{S}^{+}(p,\rho)$) *be compact and that the space be*

[6] The formulation of G (18.5) does not express this result correctly.

noncompact are both essential to produce a simple assertion. If either of these is omitted one obtains a great variety of solutions, most of which are without geometric interest. For $D \subset A^n$ see Berwald ⟨1⟩ and for $D = P^2$ see Funk ⟨2⟩; in the latter case there are also examples with symmetric distances which are not elliptic.

We conclude the discussion of desarguesian spaces by mentioning a result on Hilbert geometry. A G-space has nonpositive curvature if every point has a neighborhood $S(p, \delta)$ $(0 < \delta < \rho(p))$ such that any points a, b, c in U and the midpoints b' of a, b and c' of a, c satisfy the inequality $2b'c' \leqq bc$. (See G, Section 36.) If the space is straight then this inequality holds in the large when it holds locally (l.c.). Examples (G, Section 18) show that Hilbert geometry need not have nonpositive curvature. In fact:

(4) *The hyperbolic geometry is the only Hilbert geometry with non-positive curvature.*

This was proved by Kelly and Straus ⟨1⟩ and answers Problem (134) in G p. 406. Hilbert geometry does have nonpositive curvature in a different, for Riemann spaces equivalent, sense:

The capsules $\{x \mid xT \leqq \rho\}$ are convex for any segment T (G, Sections 18 and 41).

III. Length Preserving Maps

A map ϕ of a metric space R in which any two points can be connected by a rectifiable curve into a metric space R' *preserves length or is equilong*, if the length of any curve $x(t)$, $\alpha \leqq t \leqq \beta$, in R equals that of its image $\phi x(t)$ in R'. For $R = R'$ we speak of an equilong map of R. Local isometries are special equilong maps. An equilong map is proper if it is not an isometry.

If the metric of R is intrinsic, then a length preserving map ϕ of R is a *shrinkage*, i.e., $\phi x \phi y \leqq xy$[1]. After discussing some elementary facts concerning shrinkages, we investigate conditions under which an equilong map or local isometry is onto, although only assumed to be into.

It was observed in G(p p. 172, 173) that a cylinder with a euclidean metric does not possess proper local isometries, whereas it does if provided with a suitable hyperbolic metric. This leads to the question (Problem (27) in G, p. 405) of finding general theorems which explain this difference. *Three conditions have been found*, all applying to the euclidean cylinder (Section 10), *under which no proper local isometries exist;* they concern, respectively, mobility, the behaviour of the geodesics, and order of magnitude of volume.

Folding a piece of paper repeatedly provides examples of length preserving maps of the euclidean plane into itself. It is clear that all elementary, i.e., euclidean, hyperbolic, or spherical, spaces possess equilong maps obtained by folding or reflections in hyperplanes.

Spaces with a few proper length preserving maps abound, so that restrictions are necessary to obtain significant results. We assume that an isometry of an open sphere on another can be extended to a motion (isometry of the space onto itself) and prove that a region of injectivity, i.e., a maximal open connected set in which an equilong map is injective, *is convex. For the elementary spaces we construct all equilong maps which are locally finite* in the sense that the covering by the closures of the regions of injectivity is locally finite. Of other spaces, in particular, the hermitian and quaternion elliptic and hyperbolic spaces (which are not elementary) we prove that they do not possess proper locally finite equilong maps.

---

[1] Contraction sounds better; but this term is universally used for the special case $\phi x \phi y \leqq kxy$ with $0 < k < 1$.

9. Shrinkages, Equilong Maps, Local Isometries

A map ϕ of a set M in a metric space R into a metric space R' which need not be either continuous or single valued is called an *expansion*, a *shrinkage*, or an *isometry* if for any pair of points a, b in M and any $a' \in \phi a$, $b' \in \phi b$ respectively $a'b' \geqq ab$, $a'b' \leqq ab$ or $a'b' = ab$.

Notice that shrinkages and isometries are automatically single valued and continuous, but expansions are not. The generality is of no interest in itself but it makes (4) below a corollary of (1) and thus obviates a separate proof. Both theorems and their proofs are due to Freudenthal and Hurewicz ⟨1⟩.

(1) *Theorem. If ϕ is an expansion of the compact set $M \subset R$ into R with $M \supset \phi M$ then $M = \phi M$ and ϕ is an isometry.*

Let $a, b \in M$, put $a_0 = a$, $b_0 = b$ and define inductively a_i, b_i $(i = 1, 2, \ldots)$ in M by $a_{i+1} \in \phi a_i$, $b_{i+1} \in \phi b_i$. These points exist because $M \supset \phi M$. By hypothesis

(2) $$a_i a_k \geqq a_{i-1} a_{k-1}, \qquad b_i b_k \geqq b_{i-1} b_{k-1}, \qquad a_i b_i \geqq a_{i-1} b_{i-1}.$$

Since M is compact there is a subsequence $\{i_k\}$ of $\{i\}$ for which $\{a_{i_k}\}$ and $\{b_{i_k}\}$ converge, so that for a given $\varepsilon > 0$ and a suitable m

$$a_{i_m} a_{i_{m+1}} < \varepsilon/2 \quad \text{and} \quad b_{i_m} b_{i_{m+1}} < \varepsilon/2.$$

Putting $i_{m+1} - i_m = k > 0$ we conclude from (2) that

(3) $$a_0 a_k \leqq a_{i_m} a_{i_{m+1}} < \varepsilon/2, \qquad b_0 b_k < \varepsilon/2, \qquad \text{whence by (2)}$$
$$a_0 b_0 \leqq a_1 b_1 \leqq a_k b_k < a_0 b_0 + \varepsilon.$$

But $a_0 b_0 = a_1 b_1$ means that ϕ is an isometry. Therefore ϕM is compact. From (3) and $a_k \in \phi M$ we see that ϕM is dense in M, so $\phi M = M$.

The inverse of a shrinkage ϕ of M is a, not necessarily single valued, expansion of ϕM on M. Therefore (1) has the following corollaries:

(4) *Theorem. If ϕ is a shrinkage of $M \subset R$ into R and ϕM is compact, moreover $\phi M \supset M$ then $\phi M = M$ and ϕ is an isometry.*

(5) *If ϕ is an isometry of a compact set $M \subset R$ into R and either $M \subset \phi M$ or $\phi M \subset M$ then $\phi M = M$.*

For special spaces (4) has often interesting consequences, for example:

(6) A shrinkage ϕ of the n-dimensional spherical space S^n into itself which is not an isometry has at least one fixed point, maps at least one point on its antipode and sends at least one pair of antipodes into the same point. Moreover, ϕS^n lies in a (closed) hemisphere.

By (4) the mapping ϕ has degree 0. Therefore the first two assertions follow from well known topological theorems. The third is a consequence of the theorem of Borsuk and Ulam, see Alexandroff and Hopf [1, p. 486]. If a, a' are antipodal points mapped on the same point b and S^n has curvature 1, then $\min(ax, a'x) \leqq \pi$ for every point x, whence $b\phi x \leqq \pi/2$.

We list a few trivial frequently used observations:

(7) *If ϕ is a shrinkage of R into R' and $\phi a \phi b = ab$ then ϕ maps a segment $T(a, b)$ isometrically on a segment $T(\phi a, \phi b)$.*

For, if $c \in T(a, b)$ then

$$ab = ac + cb \geqq \phi a \phi c + \phi c \phi b \geqq \phi a \phi b = ab,$$

whence $\phi a \phi c = ac$, $\phi c \phi b = bc$. If d is a fourth point of $T(a, b)$ on the subsegment $T(c, b)$ then this argument yields $\phi c \phi d = cd$. As a corollary we have

(8) *If a, b are fixed points of the shrinkage ϕ of R into itself and $T(a, b)$ exists and is unique, then ϕ leaves $T(a, b)$ pointwise fixed.*

Thus the fixed points (if any) of a shrinkage of a euclidean or hyperbolic space into itself form a convex set.

The map ψ of an open set M in R into R' is a local isometry if every point $p \in M$ has a neighborhood $S(p, \rho_p) \subset M$ which ψ maps isometrically on $S(\psi p, \rho_p)$.

(9) *A length preserving map, hence a local isometry, of a space R with an intrinsic metric into a metric space R' is a shrinkage.*

For, given a, b in R and $\varepsilon > 0$ there is a curve C from a to b with $\lambda(C) < ab + \varepsilon$. Then

$$\phi a \phi b \leqq \lambda(\phi C) \leqq \lambda(C) = ab + \varepsilon.$$

Thus we obtain from (4) the following fact which considerably generalizes G (27.14).

(10) *Theorem. A local isometry or equilong map of a compact space with an intrinsic metric onto itself is a motion.*

In noncompact spaces the situation is different. A map of E^2 onto itself is a motion when locally isometric (since E^2 simply connected, G (27.16)), but need not be when equilong. For, if x, y are cartesian coordinates, then the map

$$x' = x, \quad y' = y + 1 \text{ for } y < 0, \ = 1 - y \text{ for } 0 \leqq y \leqq 1, \ = y - 1 \text{ for } y > 1$$

is equilong and onto but not injective.

Various statements on equilong or locally isometric maps can be strengthened, if domain invariance is assumed. As an illustration we consider two examples exhibiting phenomena which do not occur when domains are invariant.

1) For the nonnegative real axis $t \geqq 0$ with the usual distance $|t_1 - t_2|$ the map $t' = t + 1$ is an isometry which is not onto and hence, by definition, not a motion.

For the second example we define:

A map ϕ of R into R' is *locally injective* if every point p of R has a neighborhood U such that the restriction ϕ_U of ϕ to U is injective.

2) In E^2 with complex numbers as coordinates consider the compact set R consisting of 0 and the circles $z = -2 \cdot 3^{-n} + 3^{-n} e^{i\theta}$, $i = 0, 1, 2, \ldots$. As metric in R we take the intrinsic metric induced by E^2, i.e., the length of a shortest arc connecting the points along the circles.

Then the map ϕ of R into itself defined by $\phi(0) = 0$, $\phi(-2 \cdot 3^n + 3^{-n} e^{i\theta}) = -2 \cdot 3^{-n-1} + 3^{-n-1} e^{3i\theta}$ is locally injective and equilong, but not locally isometric.

Notice also that R possesses motions which leave a nonempty open set pointwise fixed without being the identity.

We call a map ϕ of R into R' *weakly open* if ϕD is open in R' whenever D is open in R and ϕ or ϕ_D map D topologically on ϕD. This weakening of the concept of an open map is needed to obtain (13) and (14) below. We need the lemma

(11) *Let R, R' be locally compact spaces with intrinsic metrics. If ϕ is a weakly open equilong map of R into R', for which the restriction ϕ_M of ϕ to the open set M is injective, then ϕ_M is a local isometry.*

For a proof choose $\delta > 0$ such that $\bar{S}(p, \delta)$ is compact and lies in M. Then ϕ_M maps $\bar{S}(p, \delta)$ topologically on $\phi \bar{S}(p, \delta) = \phi_M \bar{S}(p, \delta)$. Since ϕ is weakly open $\phi_M S(p, \delta)$ is open, so $\rho = \rho_p > 0$ exists such that $\bar{S}(p', 2\rho)$ $(p' = \phi p)$ lies in $\phi S(p, \delta)$ and is compact.

For any two points x', y' in $\bar{S}(p', \rho)$ there exists a $T(x', y') \subset \bar{S}(p, 2\rho)$. Because ϕ_M is topological on $\bar{S}(p, \delta)$ and preserves length $\phi_M^{-1} T(x', y')$ is a curve from $x = \phi_M^{-1} x'$ to $y = \phi_M^{-1} y'$ of length $x' y' \geqq x y$, so $x' y' = x y$ by (10), in particular $p x = p' x'$. Therefore $\phi_M^{-1} \bar{S}(p', \rho)$ is isometric to $\bar{S}(p, \rho)$ and contained in $\bar{S}(p, \rho)$, whence, by (5), $\phi_M^{-1} \bar{S}(p', \rho) = \bar{S}(p, \rho)$. Thus ϕ_M maps $S(p, \rho_p)$ isometrically on $\bar{S}(p', \rho_p)$ and is therefore a local isometry.

We now establish the principal result in this direction:

(12) *Theorem. If R and R' are finitely compact spaces with intrinsic metrics, then a locally injective, weakly open, equilong map of R into R' is a local isometry of R onto R'.*

By (11) there is a positive function ρ_p defined on R such that ϕ maps $\bar{S}(p, \rho_p)$ isometrically on $\bar{S}(\phi p, \rho_p)$ and hence also $S(p, \rho_p)$ on $S(\phi p, \rho_p)$ (see (1.4)) so ϕ is a local isometry of R into R'. Let $\rho(p)$ be supremum of all ε for which ϕ maps $\bar{S}(p, \varepsilon)$ isometrically on $\bar{S}(\phi \rho, \varepsilon)$. Then either $\rho(p)=\infty$ and ϕ is an isometry of R into R', or $\rho(p)<\infty$. Then $\bar{S}(p, \varepsilon) \supset \bar{S}(q, \varepsilon - pq)$ for $p\rho < \varepsilon$ shows $|\rho(p)-\rho(q)| \leqq pq$ (compare G, p. 33). Therefore $\rho(p)$ is a continuous function which has a positive minimum on any $\bar{S}(p, \delta)$ which is, by hypothesis, compact.

We must show that for a given $r' \in R'$ an $r \in R$ with $\phi r = r'$ exists. Take any $q \in R$ and let $q' = \phi q$. There is a segment T' from q' to r'. Let $\delta = \min \rho(p)$ for $p \in \bar{S}(q, q'r')$ and choose $q'_0, q', q'_1, \ldots, q'_n = r'$ on T' with $(q'_{i-1}\, q'_i\, q'_{i+1})$ and $q'_{i-1}\, q'_i < \delta$.

Because ϕ maps $\bar{S}(q, \delta)$ isometrically on $\bar{S}(q, \delta)$ there is a segment $T(q, q_1)$ which ϕ maps isometrically on the subsegment $T(q', q'_1)$ of T'. For the same reason ϕ maps a suitable segment $T(q_1, q_2)$ in $S(q_1, \delta)$ on the subsegment $T(q'_1, q'_2)$ of T'. Thus we arrive at a segment $T(q_{n-1}, q_n)$ mapped on the segment $T(q'_{n-1}, q'_n)$ of T'. With $q_n = r$ we have $\phi r = r'$.

A corollary of this theorem, (4.3) and (3.2), is

(13) *Let R, R' be G-spaces of the same finite dimension. Then a locally injective equilong map of R into R' is a local isometry of R onto R'.*

(14) *A locally injective equilong map of a finitely compact space R with an intrinsic metric and domain invariance, hence of a finite dimensional G-space, is a local isometry onto. If R is compact then ϕ is a motion.*

The last assertion follows from (10). Examples 1), 2) show that (14) does not hold without domain invariance.

10. Spaces without Proper Local Isometries

A local isometry of a noncompact G-space R onto itself is a motion if the fundamental group of R is not isomorphic to a proper subgroup (G (27.16)). Without this hypothesis the assertion may or may not be true. It holds for a cylinder with a locally euclidean metric, but is false for a cylinder with a suitable locally hyperbolic metric (G, p. 173). This leads to the problem (G, p. 405 (27)) of finding nontopological conditions under which local isometries are motions. We discuss three theorems of this type. The first is quite simple and was given by Kirk ⟨1⟩.

(1) *Theorem. A local isometry of a G-space onto itself which possess a fixed point is a motion.*

(2) *Corollary. A local isometry ϕ of a G-space on itself is a motion, if and only if, there is a motion ψ and a point p with $\psi \phi p = p$. In particular,*

a G-space with a transitive group of motions does not admit proper local isometries.

Let ϕ be a local isometry with p as fixed point. Assume ϕ is not injective, then $p_1 \neq p$ with $\phi p_1 = p$ exists because $\phi^{-1} x$ contains equally many points for all x, G (27.13). Moreover, given x, y and $\bar{x}$ with $\phi \bar{x} = x$ then $\bar{y}$ with $\phi \bar{y} = y$ and $\bar{x}\bar{y} = xy$ exists, see G (27.4, 11). We apply this to $p = x = \bar{x}$ and $p_1 = y$ and $p_2 = \bar{y}$ with $\phi p_2 = p_1$ and $p_2 p = p_1 p$. Continuing in the same way we obtain a sequence $\{p_n\}$ with $\phi p_{n+1} = p_n$ and $p_n p = p_1 p$.

If $n < m$ then $\phi^n p_n = p$, $\phi^n p_m = p_{m-n}$, and $p p_{m-n} = p p_1$ implies $p_{m-n} \neq p$, so $\phi^n p_m \neq \phi^n p_n$ and $p_n \neq p_m$. Since ϕ^n is a local isometry we have $p_i p_j \geqq 2\rho(p)$ for $i \neq j$, G (27.10), which contradicts finite compactness.

With the notation of (2), if ϕ is a motion then ψ can be chosen as ϕ^{-1}. Conversely, if a motion ψ and p with $\psi \phi p = p$ exist, then $\psi \phi$ is a local isometry, hence by (1) a motion ω, so $\phi = \psi^{-1} \omega$ is a motion.

Kirk ⟨2⟩ shows that the assertion of (1) remains valid when $\phi p = p$ is replaced by the weaker condition that $\{\phi^n p\}$ be bounded.

The second theorem is essentially due to Szenthe ⟨2, 3⟩. We present here the more general form given by Kirk ⟨3⟩. To formulate it we define a *loop at a point p* of a G-space R as a geodesic monogon L with vertex p (for the terminology see G, p. 210) such that L is the union of two segments from p to q ($q \in L$). These segments have, of course, only p and q in common.

Denote by $Q(p)$ the set of all loops at p and define

$$\lambda_i(p) = \inf_{L \in Q(p)} \lambda(L), \quad \lambda_s(p) = \sup_{L \in Q(p)} \lambda(L) \quad \text{if } Q(p) \neq \emptyset,$$

$$\lambda_i(p) = \infty, \quad \lambda_s(p) = 0 \quad \text{if } Q(p) = \emptyset.$$

Let

$$\lambda_i(R) = \inf_{p \in R} \lambda_i(p), \quad \lambda_s(R) = \sup_{p \in R} \lambda_s(p).$$

Then the following holds.

(3) *Theorem. A G-space R does not possess proper local isometries if* $\lambda_i(R) > 0$ *and* $\lambda_s(R) < \infty$.

Note that the hypotheses are satisfied in compact spaces, since $\lambda_i(p) \geqq 2\rho(p)$ and $\lambda_S(p) \leqq 2$ diameter R and also on a cylinder with a euclidean metric.

For an indirect proof of (3) assume that a proper local isometry ϕ exists. For $p_1 \in \phi^{-1} p$ let

$$W(p, p_1) = \{x \mid x \in \phi^{-1} p - p_1\}.$$

Since any two points of $\phi^{-1} p$ have at least distance $2\rho(p)$ the set $W(p, p_1)$ is closed and p_1 has a foot p_1^f on it. We claim that $\phi T \in Q(p)$ for any segment T from p_1 to p_1^f.

To see this let z_1 be the midpoint of T and $T_1 = T(p, z_1)$, $T_2 = T(z_1, p_1^f)$. There is an $r \in \phi^{-1} p$ with $r z_1 = p z$ where $z = \phi z_1$ (see the preceding proof). The definition of p_1^f and $x y \geqq \phi x \phi y$ yield

$$p_1 p_1^f = p_1 z_1 + z_1 p_1^f \geqq p_1 z_1 + z p = p_1 z_1 + z_1 r \geqq p_1 r \geqq p_1 p_1^f.$$

Therefore $p z = z_1 p_1^f = p_1 p_1^f / 2$ and

$$\lambda(\phi T_1) = \lambda(\phi T_2) = p z.$$

But ϕT_1 and ϕT_2 are geodesic curves from p to z, hence are segments, so that ϕT is a loop.

Take any point $q \in R$ and define q_n^f as the foot of $\phi^n q$ on $W(\phi^{n+1} q, \phi^n q)$. For a segment T_n from $\phi^n q$ to q_n^f the curve ϕT_n is, as just proved, a loop in $Q(\phi^{n+1} q)$ of length $\phi^n q q_n^f \leqq \lambda_s(R)$. Because ϕ^n is a local isometry there is a point $r_n \in \phi^{-n} q_n^f$ with $q r_n = \phi^n q q_n^f$. So the sequence $\{r_n\}$ is bounded and there are integers $k < l$ for which $r_k r_l < \lambda_i(R)$. We have $\phi^e r_l = q_l^f$ and

$$\phi^l q q_l^f = \phi^l r_k \phi^l r_l \leqq r_k r_l < \lambda_i(R).$$

But this would mean that $\phi T_l \in Q(\phi^{l+1} q)$ has length less than $\lambda_i(R)$.

We now turn to a third theorem, found in Busemann ⟨8⟩, which states that under certain conditions on volume a local isometry is a motion. Although the proof will show that much less suffices, we require for *a volume α in a G-space R* that α be a finitely additive measure defined on an algebra of subsets of R [2] which contains all spheres $S(p, \delta)$. If $f(\nu) \geqq 0$ is defined for $\nu = 1, 2, \ldots$ we mean, as usual, with

$$f(\nu) = o(k+1)^\nu \quad \text{that } f(\nu)(k+1)^{-\nu} \to 0 \text{ for } \nu \to \infty,$$

and prove:

(4) *Theorem. Let a volume α be defined in the G-space R such that a point q, positive numbers ε, η, and a positive integer k exist with the properties*

a) $\alpha(S(q, \rho)) > 0$ for $\rho > 0$.

b) $\varepsilon \alpha(S(q, \rho)) \leqq \alpha(S(p, \rho))$ if $\rho < \eta$ *and* $S(p, \rho)$ *is isometric to* $S(q, \rho)$.

c) $\alpha(S(q, \nu \gamma)) = o(k+1)^\nu$ *for fixed* $\gamma > 0$.

Then a local isometry ϕ of R onto itself is at most k to 1. In particular, ϕ is a motion if $k = 1$.

Assume for an indirect proof that some $k+1$ distinct points are mapped on the same point. Then $\phi^{-1} q$ also contains $k+1$ distinct points $q_1, \ldots, q_{k+1}$. Put $Q_1 = \bigcup q_j$ and connect q with q_j by a segment T_j, where

[2] For the terminology see Hewitt and Stromberg ⟨1, p.p. 4, 126⟩

$T_j = q$, if $q = q_j$ [3], $\qquad \max \lambda(T_1) = \max q q_j = \beta > 0.$

Among the points $\phi^{-1} q_j$ are the endpoints of the segments beginning at q_j and lying over T_n $(n = 1, \ldots, k+1)$. They are distinct and none have distance greater than β from q_j.

Denote the set of $(k+1)^2$ points obtained in this way by Q_2. All these points are distinct because $q_i \neq q_j$ for $i \neq j$. Define $Q_3, Q_4, \ldots$ analogously. Then Q_ν consists of $(k+1)^\nu$ distinct points and $\phi^\nu Q = q$. Because ϕ^ν is a local isometry

$$S(q', \rho(q)) \cap S(q'', \rho(q)) = \emptyset \qquad \text{for distinct } q', q'' \text{ in } Q_\nu.$$

Moreover, ϕ^ν maps $S(q', \delta)$ isometrically on $S(q, \delta)$ if $\delta = \min(\eta, \rho(q)/2)$, see G (27.10).

Writing $\alpha S(p, \rho)$ instead of $\alpha(S(p, \rho))$ we conclude from b) that

$$\varepsilon \alpha S(q, \delta) \leqq \alpha S(q', \delta) \qquad \text{for } q' \in Q_\nu.$$

For a given point r_i in Q_i there is a point r_{i-1} in Q_{i-1} $(Q_0 = \{q\})$ with $r_i r_{i-1} \leqq \beta$. Applying this to $r_\nu = q'$ we obtain points $r_{\nu-1}, \ldots, r_1, r_0 = q$ with

$$q' q \leqq \sum_{i=1}^{\nu} r_i r_{i-1} \leqq \nu \beta, \qquad \text{hence } S(q', \delta) \subset S(q, \nu\beta + \delta).$$

Because the $S(q', \delta)$ are disjoint

$$\sum_{q' \in Q_\nu} \alpha S(q', \delta) \leqq \alpha S(q, \nu\beta + \delta)$$

and for $\nu \geqq \delta/\beta$

$$(k+1)^\nu \varepsilon \alpha S(q, \delta) \leqq \alpha S(q, 2\nu\beta),$$

which contradicts c) for $\gamma = 2\beta$.

We apply (4) to a locally euclidean space R_E with its ordinary volume α_E. For any point q and $\rho \leqq \rho(q)/2$ the sphere $S(q, \rho)$ is isometric to a sphere in E^n, so that

$$\alpha_E(S(q, \rho)) = \pi_n \rho^n, \qquad \text{where } \pi_n = \pi^{n/2}/\Gamma(1 + n/2),$$

and any $S(p, \rho)$ isometric to $S(q, \rho)$ has the same volume. Therefore the hypotheses of (4) are satisfied for each q with $\varepsilon = 1$, $\eta = \rho(q)/2$ and $k = 1$ because for any γ and ν

$$\alpha_E(S(q, \nu\gamma)) \leqq \pi_n (\nu\gamma)^n = o(2^\nu).$$

Thus we find using (9.14):

(5) *A local isometry of a locally euclidean space into itself is a motion.*

[3] Using (1) we could conclude that no $q_j = q$.

This result can be generalized. The *n-dimensional Hausdorff measure* (see Hausdorff ⟨1⟩) is defined for any set M in a metric space with a countable base as follows:

Denote the diameter of the set N by $\delta(N)$. For a given $\varepsilon>0$ consider all decompositions $\{M_i\}$ of M into a countable number of sets M_i with $\delta(M_i)<\varepsilon$. Form

$$\alpha_\varepsilon(M)=\inf_{\{M_i\}}\pi_n 2^{-n}\sum_i \delta^n(M_i).$$

Then $\alpha_\varepsilon(M)\leqq\alpha_{\varepsilon'}(M)$ for $\varepsilon>\varepsilon'$. Therefore

$$\alpha_n(M)=\lim_{\varepsilon\to 0+}\alpha_\varepsilon(M)$$

exists and clearly $\alpha_n(M)=\alpha_n(M')$ for isometric M, M'. The number $\alpha_n(M)$ is the n-dimensional outer Hausdorff measure. It defines a possibly trivial measure on all Borel sets, which coincides with ordinary volume or Lesbegue measure of R in an n-dimensional locally euclidean space.

Consider a continuous map ψ of R into R' satisfying

$$x\,y\leqq\beta\,\psi x\,\psi y \quad \text{for some fixed } \beta>0.$$

Then ψ is a homeomorphism and for any set M in R

$$\alpha_n(M)\leqq\beta^n\alpha_n(\psi M).$$

Therefore

(6) *Theorem. If the G-space R can be mapped into an n-dimensional locally euclidean space R_E by a continuous map ψ satisfying $x\,y\leqq\beta\,\psi\,x\,\psi\,y$, and if for some $q\in R$ the n-dimensional Hausdorff measure of the spheres $S(q,\delta)$ in R is positive, then a local isometry of R into itself is a motion.*

Because ψ is a homeomorphism, $\dim R\leqq n$ and the local isometry is onto by (9.14).

From $\alpha_n(M)\leqq\beta^n\alpha_E(\psi M)$ and $\psi S(p,\rho)\subset S(\psi p,\beta\rho)$ we conclude

$$\alpha_n(S(p,\rho))\leqq\beta^{2n}\pi_n\rho^n \quad \text{for any } p,\rho,$$

so that c) in (4) holds with $k=2$. Since the Hausdorff measures of isometric sets are equal b) holds too, and a) is a hypothesis. Notice that a) is satisfied if $\dim R=n$. This follows from the connection between dimension and Hausdorff measure, see Hurewicz and Wallman [1, Chapter VII].

If R_m is a locally Minkowskian space with distance $m(x,y)$ then R_m carries a locally euclidean distance $E(x,y)$ such that

$$\beta_1 E(x,y)\leqq m(x,y)\leqq\beta_2 E(x,y)$$

with suitable $\beta_i>0$, see G, p. 192. Therefore

(7) *Locally Minkowskian spaces do not admit proper local isometries and may be substituted for R_E in (6).*

11. Proper Equilong Maps

A discussion of all spaces which admit proper equilong maps is neither feasible nor interesting, because there are many spaces with isolated proper length preserving maps like an ellipsoid with different axes. There may, however, be a general theorem to the effect that such a map can exist only when the space possesses a local reflection in some nowhere dense convex set H which locally separates the space (compare (8) below). The methods hitherto developed do not permit us to tackle the question in this generality.

Significant results can be obtained for G-spaces with domain invariance (DI) and the following extension property

E: *An isometry ϕ of a sphere $S(p, \rho)$ in R on a sphere $S(q, \rho)$ $(\rho > 0)$ in R can be extended to a motion of R*[4].

The elementary and the elliptic spaces have this property and so do all simply connected Riemann spaces with analytic metrics. Among the latter we mention for later purposes the following, which do not have constant curvature, but do possess pairwise transitive groups of motions.

(*) *The hermitian elliptic and hyperbolic spaces of (real) dimension greather than 2. The quaternion elliptic and hyperbolic spaces of dimension greater than 4. The elliptic and hyperbolic Cayley planes.*

These are besides the elementary and elliptic spaces the only spaces enjoying pairwise mobility[5].

Minkowskian spaces have property E because of the existence of dilations.

Finally, E is trivially satisfied in the generic case of a *totally inhomogeneous* space R. This means that for any p and $\rho > 0$ the only isometric map of $S(p, \rho)$ into R is the *inclusion* (the restriction of the identity to $S(p, \rho)$).

As before we call a length preserving map of R into itself an equilong map of R. First a lemma:

(1) *If the G-space R has the properties DI and E and the equilong map ϕ of R is injective on the connected open set M, then ϕ_M can be extended to a motion of R.*

By (9.11) and the connectedness of M there is a sequence of points $p_1, p_2, \ldots$ and of positive numbers $\rho_1, \rho_2, \ldots$ such that ϕ maps $S(p_i, \rho_i) \subset M$

[4] The material of Section 9 and 11 is taken from Busemann ⟨9⟩.

[5] See G Sections 53 – 55 for and Section 19 here.

isometrically on $S(\phi p_i, \rho_i)$, $M = \bigcup_i S(p_i, \rho_i)$ and

$$S(p_{i+1}, \rho_{i+1}) \cap S_i \neq \emptyset \quad \text{where } S_i = \bigcup_{k=1}^{i} S(p_k, \rho_k).$$

By hypothesis the restriction of ϕ to $S(p_i, \rho_i)$ can be extended to a motion β_i of R. We prove $\phi = \beta_1$ for each S_i.

This is trivial for S_1. Assume it has been proved for S_n. Then $\beta_{n+1} = \phi = \beta_1$ on $S(p_{n+1}, \rho_{n+1}) \cap S_n \neq \emptyset$. Therefore $\beta_{n+1} = \beta_1$ (see G (28.8)), in particular $\beta_{n+1} = \beta_1 \phi$ on S_{n+1}.

A *region of injectivity* of an equilong map ϕ of R is a maximal open connected set on which ϕ is injective. These regions will prove basic for our study. In contrast to fundamental sets they are uniquely determined when DI and E hold. This is part of the following long theorem which contains all the pertinent information on this subject.

(2) *Theorem. Let the G-space R have properties DI and E and let D be a region of injectivity of the equilong map ϕ of R.*

Then D contains with any two points x, y all T(x, y). Hence, if x, y lie in $\bar{D}$ then at least one $T(x, y) \subset \bar{D}$. Moreover $\phi_{\bar{D}}$ is injective and the restriction of a motion to $\bar{D}$.

No boundary point of D has a neighborhood on which ϕ is injective. Different regions of injectivity are disjoint.

By (1) there is a motion β with $\beta x = \phi_D x = \phi x$ for $x \in D$. Therefore $\beta^{-1}\phi$ is an equilong map of D which leaves D pointwise fixed. Denote by F the set of all fixed points of $\beta^{-1}\phi$. Then F is closed, contains D and $\phi x = \beta x$ on F so ϕ_F is injective.

Denote the interior of F by F_i and consider a segment T from x to y $(x \neq y)$ with $x \in F_i$ and $y \in F$. Let $S(x, \rho) \subset F_i$, $\rho < xy$. For $z \in T$ and $0 < xz < \rho$ the segment $T(z, y)$ is unique, hence lies by (9.8) in F, so $T \subset F$.

If x, y lie in F_i then $T \subset F_i$. For, assume $T \cap (F - F_i) \neq \emptyset$ and let z be the first point coming from x in this intersection. Choose u, v on T such that x, u, z, v lie in this order on T and $uv < \rho(v)/3$, then ρ with $0 < \rho < \rho(v)/3$ and $S = S(u, \rho) \subset F_i$. For $w \in S(u, \rho)$ the segment $T(w, v)$ is unique because $w, v < \rho(v)$, moreover $T(w, v) \subset F$. Therefore

$$W = \bigcup_{w \in S} T(w, v) - \{v\} \subset F.$$

But W is open because the $T(w, v)$ are unique and $z \in W \subset F$. This would imply $z \in F_i$.

Thus F_i is connected, contains D and ϕ is injective on F_i. The maximality of D implies $F_i = D$.

Let ϕ be injective on $S' = S(p, \rho)$ with $S' \cap F \neq \emptyset$. Because of (1) ϕ coincides on S' with a motion β' and $\beta' = \beta$ since this holds on $S \cap F_i \neq \emptyset$.

The definition of F yields $S(p,\rho)\subset F$. This implies the last two assertions in (2).

We call an equilong ϕ of R *locally finite*, if R is the union of the closures of the regions of injectivity of ϕ and if this covering is locally finite. Under the hypotheses of (2), if ϕ is locally finite, there will be a countable number of regions of injectivity $D_1, D_2, \ldots$ and as in the proof of (2) we may assume that ϕ leaves one of the D_i pointwise fixed. Unless stated otherwise we choose the notation so that D_1 stays pointwise fixed. There is more than one D_i if ϕ is proper.

Our aim is to decide for some important classes of spaces whether they admit proper locally finite (p.l.f.) equilong maps and, if so, to construct them.

(3) *A totally inhomogeneous space with domain invariance admits no other locally finite equilong maps than the identity.*

This follows at once from (2) since the identity is the only motion[6].

Next we consider the trivial, but important, *one-dimensional* cases, the straight line and a great circle. We realize the first case as the x-axis. The region of injectivity of a p.l.f. equilong map are intervals or rays whose endpoints form a discrete set. Give any D_i of this type which are disjoint and whose closures cover the real axis. We may assume that D_1 has a right endpoint x_2. Let $D_2', D_3', \ldots$ be the regions to the right of D_1 in this order and β_i the reflection of the real axis in the left endpoint x_i of D_i'. If β_1 is the identity and $D_1'=D_1$ the equilong map ϕ which leaves D_1 pointwise fixed, and has the D_i as regions of injectivity is unique and is given for $x\in D_1'\cup D_2' \ldots$ by

$$\phi x=\beta_1\beta_2,\ldots,\beta_j x \quad \text{if } x\in D_j'.$$

The procedure is analogous for the D_i to the left of D_1 (if any).

A great circle may be realized as a circle C in E^2 with the length of the shorter arc as distance. Orient C. A p.l.f. equilong map ϕ has a finite number of regions of injectivity which are arcs of C, say $D_1, \ldots, D_m$, in the order of the orientation. If β_1 is the identity and β_i $(i>1)$ is the reflection of C in the diameter passing through the initial point of D_i then ϕ is again given by

(4) $$\phi x=\beta_1\ldots\beta_j \quad \text{if } x\in D_j,$$

but two conditions must be satisfied: m must be even, otherwise ϕ would be injective on $D_m\cup D_1$, and the endpoint of D_m must stay fixed. In terms of the lengths δ_i of D_i this means

(5) $$\delta_1+\delta_3\cdots+\delta_{m-1}=\delta_2+\delta_4+\cdots+\delta_m \quad (m \text{ even}).$$

[6] Assuming the latter is not sufficient. It is easy to metrize the plane as a (Riemannian) G-space which has p.l.f. equilong maps and whose only motion is the identity.

For a later application we notice: if β_{m+1} is the reflection in the diameter through the initial point of D_1 then

(6) $$\beta_1 \beta_2 \dots \beta_{m+1} = \beta_1 = 1.$$

Conversely, if $D_1, \dots, D_m$ are disjoint open subarcs of C with $\bigcup \bar{D}_i = C$ and following each other in the given order, which satisfy (5), then (4) defines an equilong map of C with the D_i as regions of injectivity which leaves D_1 pointwise fixed.

Next we show

(7) *A Minkowskian G-space R* (dim $R = n \geqq 2$) *admits proper locally finite equilong maps if, and only if, it possesses the reflection in some hyperplane.*

Assume the reflection in the hyperplane H exists. Then the reflections in the hyperplanes parallel to U also exist. Choose affine coordinates $x^1, \dots, x^n$ such that it is given by $x^n = 0$ and the x^n-axis is perpendicular to H in the Minkowski sense. If $x^n \to \psi x^n$ defines a p.l.f. equilong map of the x^n-axis into itself, then the map ϕ given by

$$\phi(x^1, \dots, x^n) = (x^1, \dots, x^{n-1}, \psi x^n)$$

is evidently a p.l.f. equilong map of R. The construction of all p.l.f. equilong maps of a Minkowski space is implicitly contained in (10).

Conversely, let R possess a p.l.f. equilong map ϕ. Its regions of injectivity ($\phi x = x$ on D_1) are convex polyhedral domains and there are at least two of them. Therefore D_1 has a boundary point p such that for suitable $\rho > 0$ the sphere $S(p, \rho)$ intersects $\bar{D}_1$ and a single other $\bar{D}_i$, say $\bar{D}_2$. Then $S = S(p, \rho) \cap \bar{D}_1 \cap \bar{D}_2$ lies in a common $(n-1)$-face of $\bar{D}_1$ and $\bar{D}_2$ and ϕ coincides on $S(p, \rho) \cap \bar{D}_2$ with a motion β of R, see (2). Since β leaves S pointwise fixed it is either the identity or the reflection in the hyperplane containing S. The first case is impossible, because D_1 would then not be a region of injectivity.

Consider a G-space R satisfying E which is a Riemann or Finsler space and admits a p.l.f. equilong map ϕ. As above there are D_1, D_2 and $p \in \bar{D}_1 \cap \bar{D}_2$, $S(p, \rho) \subset \bar{D}_1 \cup \bar{D}_2$. Choose ρ so small that $T(x, y)$ is unique in $S(p, \rho)$ and $S(p, \rho)$ is convex (see, for example, G, p. 162). Then $S = S(p, \rho) \cap \bar{D}_1 \cap \bar{D}_2$ is also convex and separates S. If dim $R = n$ then dim $S = n-1$ and S is, in the language of differential geometry, a totally geodesic set. Thus:

(8) *An n-dimensional Finsler space which is a G-space with property E and admits a proper locally finite equilong map must contain some $(n-1)$-dimensional totally geodesic set.*

The spaces listed under (*) do not contain such sets. This can be seen as follows: If S is an $(n-1)$-dimensional totally geodesic set in a space with a pairwise transitive group of motions, let (L, p) be a lineal element normal to S at $p \in S$. By pairwise transitivity (L, p) can be moved into a given lineal element (L', p') whereby S goes into a totally geodesic set S' normal to (L', p') at p'. It follows from Beltrami's Theorem (see G Section 15, also Cartan [2, Chapter V]) that the space has constant curvature.

(9) *Theorem. The hermitian elliptic and hyperbolic spaces of dimension greater than 2, the quaternion elliptic and hyperbolic spaces of dimension greater than 4, and elliptic and hyperbolic Cayley planes do not admit proper locally finite equilong maps.*

It is probable, but the present method does not prove, that these spaces do not admit *any* proper equilong maps.

Let R be an elementary space of dimension greater than 1, and $D_1, D_2, \ldots$ the regions of injectivity of a p.l.f. equilong map ϕ of R which leaves D_1 pointwise fixed. The r-faces of $\{D_i\}$ are the r-faces of the individual D_i.

A *string* $s=(D'_1, \ldots, D'_r)$ of D_i is defined by the property that D'_i and D'_{i+1} have a common $(n-1)$-face. Hence they are either identical then we put $\beta_{i+1}=1$ or they have precisely one common $(n-1)$-face and β_{i+1} is the reflection of R in the hyperplane containing the face. With $\beta_1=1$ we put

$$\beta(s)=\beta_1 \ldots \beta_r.$$

The discussion of the Minkowski case shows that ϕ is given by

(10) $\phi x=\beta(s)x$, *if* $x \in D_j$ *and* s *is a string of the form* $s=(D'_1=D_1, D'_2, \ldots, D'_r=D_j)$.

Conversely, if for a given locally finite division of an elementary space into convex polyhedral regions $D_1, D_2, \ldots$ the relation (10) defines a map, i.e., if ϕ is independent of the string leading from D_1 to D_j, then ϕ is a p.l.f. equilong map with the D_i as regions of injectivity.

To find conditions for this independence consider an $(n-2)$-face f of $\{D_i\}$ (if any) and let $D^f_1, \ldots, D^f_{m_f}$ be the D_i which have f as face in cyclical order. D^f_1 is one string from D^f_1 to D^f_1 and $(D^f_1, D^f_2, \ldots, D^f_{m_f} D^f_1)$ is a second. If β_{m_f+1} is the reflection in the hyperplane containing the common $(n-1)$-face of $D^f_{m_f}$ and D^f_1 then we must have

$$\beta_1 \beta_2 \ldots \beta_{m_f+1}=\beta_1=1.$$

If δ_i denotes the dyhedral angle of D_i at f then the discussion of the case of the cricle C shows that this condition is equivalent to

$$m_f \quad \text{is even and} \quad \delta_1+\delta_3+\cdots+\delta_{m_f-1}=\delta_2+\delta_4+\cdots+\delta_{m_f}.$$

Surprisingly, requiring this for every $(n-2)$-face proves also sufficient for (10) to define a map. The proof can be found in Busemann ⟨9⟩.

It is omitted here because the details are cumbersome without offering serious difficulties. Thus we arrive at the following very appealing result:

(11) *Theorem. Let* $\{D_i\}$ *be a locally finite division of the elementary space* R ($\dim R \geqq 2$) *into at least two convex regions. For a given* $(n-2)$-*face* f *of* $\{D_i\}$ *let* $D_1^f, \ldots, D_{m_f}^f$ *be the* D_i *having* f *as face in cyclical order and* δ_i *the dyhedral angle of* D_i^f *at* f.

The D_i *form the regions of injectivity of a proper equilong map of* R *if, and only if, for each* $(n-2)$-*face* f *the number* m_f *is even and*

$$\delta_1+\delta_3+\cdots+\delta_{m_f-1}=\delta_2+\delta_4+\cdots+\delta_{m_f}.$$

The equilong map with the D_i *as regions of injectivity which leaves* D_1 *pointwise fixed is given by* (10).

IV. Geodesics

This chapter deals with the behaviour of geodesics in various situations. We begin with a remarkable theorem on closed hyperbolic space forms which generalizes to n dimensions a theorem which even in the Riemannian case was known only for two, assuming neither symmetry of the distance nor, more surprisingly, local prolongability of segments or uniqueness of prolongation. The method of proof is due to Efremovič and Tihomirova ⟨1⟩ who invented it for a different purpose.

If the universal covering space R of a G-space R is related to R by a locally isometric map Ω, then the fundamental group of R can be realized as the group Δ of motions ψ of $\bar{R}$ which lie over the identity of R, i.e., $\Omega\psi = 1_R$, see G, Sections 27, 28. For straight $\bar{R}$ it is shown in G that the lines carried into themselves by motions of Δ, which we call axes of these motions, provide much information on closed geodesics in R.

It turns out that axes often exist when $\bar{R}$ is not straight (or conjugate points occur in R) and that their principal properties remain valid. Section 13 treats axes and extends with frequently much simpler proofs results established by Morse, Hedland and others for Riemannian metrics. Several theorems are also new in this case.

Our next topic (Sections 14, 15) is of an entirely different character in that it does not generalize a Riemannian situation. As such it is of particular interest for Finsler and hence G-spaces. We ask, which conditions a system Σ of curves on a surface and a group Γ (which may be trivial) of collineations of Σ must satisfy, so that *the surface can be metrized as a G-space with the curves in Σ as geodesics and Γ as groups of motions.* We will solve this problem in many significant cases beyond those treated in G, using an idea from integral geometry.

In the Riemannian case even the corresponding local problem (with $\Gamma = \{1\}$) is unsolved, and if a solution exists it is unique up to a factor, with few exceptions (Liouville surfaces, see Darboux ⟨1⟩, Sections 600–602), so that the problem in the large is not interesting. In our case the global problem, if it has a solution, has for $\dim \Gamma < 2$ so many that the well posed question concerns the topological properties of Σ rather than the determination of all metrics.

Next *we ask whether* for everywhere continuously differentiable and regular G-surfaces (Section 5) or Finsler surfaces of class C^∞ *a true analogue to the Gauss-Bonnet Theorem exists*, which means this (details in Section 16): is there an angular measure which 1) depends only on the normal tangent space and such that 2) the excess function on geodesic polygonal regions induced by this angle can always be extended to a completely additive set function on the Borel sets? We give a negative answer by showing that a quasi-hyperbolic plane possesses a unique angle satisfying 2) and that this angle does not have property 1).

In Section 17 we discuss different aspects of conjugacy, in particular, conjugacy to points at infinity, which was first investigated by Nasu ⟨1, 2, 3⟩ and deserves further study.

12. Closed Hyperbolic Space Forms

The principal goal of this section is Theorem (2) which states that on a closed hyperbolic space form with an intrinsic, not necessarily symmetric, distance there is a class of geodesics and half geodesics which behave uniformly like the hyperbolic geodesics and half geodesics. For surfaces with Riemann metrics this was proved by Morse [1] and for the case of intrinsic metrics satisfying the axioms P of the local existence and U of the uniqueness of prolongation (see Section 2) by Zaustinsky ⟨2⟩. The general result is a simple consequence of Theorem 1 below, which Busemann ⟨10⟩ obtained by reinterpreting a method developed by Efremovič and Tihomirova ⟨1⟩.

We realize the n-dimensional hyperbolic space H^n of curvature -1 by its Klein model, so that H^n appears as the interior of the units sphere K in E^n and a hyperbolic line is an open euclidean segment with endpoints p_∞, q_∞ on K. We denote this line by $L_h(p_\infty, q_\infty)$. Similarly, a hyperbolic ray with origin a in H^n and endpoint p_∞ on K is denoted by $R_h(a, p_\infty)$. With $T_h(a, b)$ we mean the hyperbolic segment and with $U_h(M, \rho)$ the hyperbolic ρ-neighborhood of the set M in H^n.

Assume that H^n is provided in addition with a not necessarily symmetric intrinsic distance xy. If $x(t)$ $(-\infty < t < \infty)$ represents a line (= straight line) with respect to xy and $\lim_{t\to-\infty} x(t)$ and $\lim_{t\to\infty} x(t)$ exist, then they are necessarily points of K. We call them the initial and endpoint of the line to which we refer as an $L(p_\infty, q_\infty)$.[1] A receding (progressing) ray is defined as a curve $x(t)$ with $-\infty < t \leqq 0$ $(0 \leqq t < \infty)$ and $x(t_1)x(t_2) = t_2 - t_1$ for $t_1 < t_2$. If $p_\infty = \lim_{t\to\pm\infty} x(t)$ exists; we call it the initial (end) point of the ray which we denote by $R(p_\infty, x(0))$ (resp. $R(x(0), p_\infty)$). The principal theorem is this:

(1) *Theorem. Let xy be an intrinsic metric in H^n which is topologically equivalent to $h(x, y)$ and let positive numbers $\alpha \leqq \alpha'$ exist such*

[1] Convergence of $x(t)$ refers to E^n. The point p_∞, q_∞ are not part of the set.

that

$$xy<\alpha \quad \textit{implies} \quad h(x,y)<\alpha' \quad \textit{and} \quad h(x,y)<\alpha \quad \textit{implies} \quad xy<\alpha'.$$

Then for given a in H^n and given $p_\infty \neq q_\infty$ on K there exists a (not necessarily unique) progressing ray $R(a, p_\infty)$, receding ray $R(p_\infty, a)$ and a line $L(p_\infty, q_\infty)$.

Each progressing ray has an endpoint, each receding ray has an initial point, and each line has an initial point and an endpoint distinct from the initial point.

There is a positive γ such that for any a in H^n and any $p_\infty \neq q_\infty$ on K

$$R(p_\infty, a) \cup R(a, p_\infty) \subset U_h(R_h(a, p_\infty), \gamma)$$

and

$$L(p_\infty, q_\infty) \subset U_h(L_h(p_\infty, q_\infty), \gamma).$$

The emphasis lies, of course, on the fact that γ *is independent* of a, p_∞, q_∞. Also, the inclusion holds for all $R(a, p_\infty)$ if a and p_∞ do not determine the ray uniquely, similarly for $R(p_\infty, a)$ and $L(p_\infty, q_\infty)$.

The proof is long, not only because the theorems has many parts, but some parts, in particular the existence of γ, require intricate estimates. These are essentially due to Efremovič and Tihomirova ⟨1⟩. Attempts to simplify them materially failed. The proof is decomposed into several steps.

(a) *There is a $\beta>0$ such that $xy>\alpha$ implies $h(x,y)<\beta xy$ and $h(x,y)>\alpha$ implies $xy<\beta h(x,y)$.*

Let $xy>\alpha$ and $(m-1)\alpha<xy\leqq m\alpha$ so that $m\geqq 2$. There is a curve $z(t)\,(0\leqq t\leqq\delta<xy+\alpha)$, where t is arc-length, with $\lambda(x)=\delta$, $z(0)=x$, $z(\delta)=y$. Choose $t_0=0<t_1<\cdots<t_{m+1}=\delta$ with $t_i-t_{i-1}<\alpha$. Putting $x(t_i)=x_i$ we have $x_{i-1}x_i\leqq t_i-t_{i-1}$, hence

$$h(x,y)\leqq\sum h(x_{i-1},x_i)<(m+1)\,\alpha'=\frac{(m+1)\,\alpha'}{(m-1)\,\alpha}(m-1)\,\alpha<\frac{3\alpha'}{\alpha}xy.$$

The second part is even simpler since $x(t)$ can be chosen as *a* $T_h(x,y)$.

Since the hyperbolic balls are compact we conclude from (a) and (1.5):

(b) *The balls $\bar{S}^{\pm}(p,\rho)$ with respect to xy are compact. Hence $T(a,b)$ exists for given $a\neq b$.*

It also follows that $xy\leqq\alpha$ implies $h(x,y)\leqq\alpha'$ etc.

(c) *For a given $\varepsilon>0$ there is an $M(\varepsilon)$ such that for $ay+yb=ab$ and $ay>M(\varepsilon)$ or $by+ya=ba$ and $ya>M(\varepsilon)$ the hyperbolic angle yab is less than ε.*

It suffices to consider the first case. First choose M_0 such that $h(a,y) \geqq M_0$ and $h(y,b) \leqq \alpha'$ imply $yab < \varepsilon$. Then $ay \geqq \beta M_0$ and $yb \leqq \alpha$ imply $yab < \varepsilon$.

Let $ay \geqq \beta M_0$ and $yb > \alpha$, then $h(a,y) \geqq M_0$ and $h(y,b) > \alpha'$. With $\alpha \leqq \alpha_0 < 2\alpha$ let $yb = m\alpha_0$ and choose on a segment $T = T(y,b)$ points $x_0 = y, x_1, \ldots, x_m = b$ with $x_{i-1}x_i = \alpha_0$. Then

$$2\beta\alpha > \beta\alpha_0 \geqq h(x_{i-1}, x_i)$$

and for $z \in T_h(x_{i-1}, x_i)$

$$h(a,z) \geqq h(a,x_i) - h(z,x_i) \geqq h(a,x_i) - 2\beta\alpha.$$

Let Σ' be the hyperbolic sphere of curvature 1 about a and for any $x \neq a$ let x' be the hyperbolic projection of x from a on Σ'. For any two points w_1', w_2' of Σ' denote $\delta(w_1', w_2')$ their intrinsic distance on Σ' so that $\delta(w_1', w_2')$ equals the angle $w_1'\, a\, w_2'$. If dw'^2 is the line element of Σ' and $h(a,x) = r$, then the hyperbolic line element is

$$ds_h^2 = dr^2 + \sinh^2 r\, dw'^2.$$

Therefore

$$\begin{aligned} h(x_{i-1}, x_i) &\geqq \delta(x_{i-1}', x_i') \sinh(h(a,x_i) - 2\beta\alpha) \\ &\geqq \delta(x_{i-1}', x_i') \sinh(\beta^{-1} a x_i - 2\beta\alpha) \end{aligned}$$

since $h(x_{i-1}, x_i) \leqq 2\beta\alpha$ we find

$$\delta(y', b') \leqq \sum \delta(x_{i-1}', x_i') \leqq 2\beta\alpha \sum_{i=0}^{\infty} \cosh[\beta^{-1}(ay + i\alpha) - 2\beta\alpha],$$

provided that $ay > 2\beta^2\alpha$. The right side will be less than ε when ay is large. This proves (c).

It follows that each progressing ray has an endpoint and each receding ray has an initial point, also, that a line has an initial and an endpoint, but not yet that these are different.

It is also easy to deduce from (c) that for given a and p_∞ rays $R(a, p_\infty)$ and $R(p_\infty, a)$ exist. For on $R_h(a, p_\infty)$ take x_i with $h(a, x_i) = i$. A sequence of segments $T(a, x_i)$ contains subsequences which converge to a ray, precisely: if $z_i(t)$ $(0 \leqq t \leqq a x_i)$ represents $T(a, x_i)$ then for a subsequence $\{z_\nu(t)\}$ of $\{z_i(t)\}$ the points $z_\nu(t)$ tend for each t to a limit $z(t)$ and $z(t)$ represent a ray. Because $q_\infty = \lim_{i \to \infty} z(t)$ exists, (c) guarantees (with the previous notation) the existence of a t_0 such that $\delta(z'(t_0), q_\infty') < \varepsilon/3$. Also, for a suitable $N > t_0$

$$\delta(z_\nu'(t_0), p_\infty') = \delta(z_\nu'(t_0), x_\nu') < \varepsilon/3$$

and

$$\delta(z_\nu'(t_0), z'(t_0)) < \varepsilon/3 \quad \text{if} \quad \nu > N.$$

So

$$\delta(p_\infty', q_\infty') < \varepsilon \quad \text{and} \quad p_\infty = q_\infty.$$

The next step is a trivial estimate needed in the crucial argument.

(d) *Let $T=T(a,b)$ be a segment of length $ab>\alpha$ and let $x_0=a, x_1, \ldots, x_k=b$ lie in this order on T with $x_{i-1}x_i \geqq \alpha$. Then*

$$\sum_{i=1}^{k} h(x_{i-1}, x_i) \leqq \beta^2 h(a,b).$$

This follows from

$$\sum h(x_{i-1}, x_i) \leqq \beta \sum x_{i-1} x_i = \beta\, ab \leqq \beta^2 h(a,b).$$

We come to the existence of γ for rays and take the case of a progressing ray $R=R(a,p_\infty)$. In H^n we introduce a system of cylindrical coordinates ζ, ρ, w with $R_h(a,p_\infty)$ as nonnegative ζ-axis. Let Σ be the $(n-2)$-sphere of curvature 1 about a in the hyperplane normal to $R_h(a,p_\infty)$ at a. For any point x we define ζ as the abscissa of its hyperbolic foot x_f on the ζ-axis, $\rho=h(x,x_f)$ and w is for $\rho \neq 0$ the point where the half plane bounded by the ζ-axis through z intersects Σ. The hyperbolic line element is then

(e) $ds_h^2 = d\rho^2 + \cosh^2\rho\, d\zeta^2 + \sinh^2\rho\, dw^2$, so $ds_h \geqq \cosh\rho\, |d\zeta|$.

Let $z(t)=(\zeta(t), \rho(t), w(t))$ $(t \geqq 0)$ represent R. We want to show that

(f) $\rho(t) < \delta(4\beta^2+1) + \alpha\beta = \gamma'$, *where* $\delta = \text{area}\cosh 2\beta^2 + 2\alpha\beta$.

First we show that $\rho(t)\to\infty$ for $t\to\infty$ is impossible. If this were the case then a sequence $\{t_k\}$ with $t_{k+1}-t_k>\alpha$, $\rho(t_k)=k$ and $\rho(t)>k$ for $t>t_k$ would exist.

Put $z(t_k)=z_k=(\zeta_k, \rho_k, w_k)$ and on a $T(z_k, z_{k+1})$ choose points x_i as in (d) with $\alpha \leqq x_{i-1}x_i < 2\alpha$. Then

$$\beta^{-2} \leqq \frac{h(z_k, z_{k+1})}{\sum h(x_{i-1}, x_i)} \leqq \frac{k + |\zeta_{k+1}-\zeta_k| + k + 1}{\sum h(x_{i-1}, x_i)}.$$

Since $\rho(t) \geqq k$ for $t \geqq k$ we have $\rho > k - 2\alpha\beta$ for $x=(\zeta,\rho,w) \in T_h(x_{i-1}, x_i)$,[2] hence by (e)

$$\sum h(x_{i-1}, x_i) > |\zeta_{k+1}-\zeta_k| \cosh(k-2\alpha).$$

Combining the last two estimates yields

$$|\zeta_{k+1}-\zeta_k| < \frac{2k+1}{\beta^{-2}\cosh(k-2\alpha\beta)-1} = \varepsilon_k.$$

But $\sum^{\infty} \varepsilon_k$ converges, so $|\zeta_k|$ would be bounded, which contradicts (c).

[2] This follows from $h(x_{i-1}, x_i) < 2\alpha\beta$. If $x \in T_h(x_{i-1}, x_i)$ and $h(x, x_i) = \min(h(x, x_{i-1}), h(x, x_i))$ then

$$\alpha\beta + \rho + \alpha\beta > h(x_i, x) + h(x, x_f) + h(x_f, x_{if} f) \geqq h(x_i, x_{if}) \geqq k.$$

The second step consists in proving the existence of a sequence $t'_k \to \infty$ with $\rho(z(t'_k)) \leqq \delta$. If this were false, then $\rho(t) > \delta$ for $t \geqq t_0$ with suitable t_0. Since $\rho(t)$ does not tend to ∞ there is a δ' and a sequence $t_k \geqq t_0$ with $t_{k+1} - t_k > \alpha$ for which

$$\delta' > \rho_k = \rho(t_k) > \delta.$$

With analogous notations as above we conclude from $\zeta(t) \to \infty$ the existence of $k_1 < k_2$ with

$$\text{(g)} \qquad \eta = |\zeta_{k_2} - \zeta_{k_1}| > 2\delta'.$$

On $T(z_{k_1}, z_{k_2})$ choose x_i as before. Then $\rho(t) > \delta - 2\alpha\beta^2$ on $T_h(x_{i-1}, x_i)$ and hence (see (f))

$$\sum h(x_{i-1}, x_i) > \cosh(\delta - 2\alpha\beta)\, 2\eta = 2\beta^2 \eta.$$

This would yield

$$\frac{h(z_{k_1}, z_{k_2})}{\sum h(x_{i-1}, x_i)} < \frac{2\delta' + \eta}{2\beta^2 \eta} < \frac{1}{2\beta^2} + \frac{1}{2\beta^2} = \frac{1}{\beta^2}$$

in contradiction to (d).

This consideration also shows

(h) *If* $\rho(t) \geqq \delta$ *for* $t' \leqq t \leqq t''$, $t'' - t' > \alpha$ *and* $\rho(t') = \rho(t'') = \delta$ *then* $|\zeta(t'') - \zeta(t')| \leqq 2\delta$.

For, if $\eta = |\zeta(t'') - \zeta(t')| > 2\delta$, then the argument following (g) applies without any change with δ replacing δ'.

Under the hypothesis of (h) let $\rho_M = \max\limits_{t' \leqq t \leqq t''} \rho(t)$. Either $\rho(t) = \delta$ on $[t', t'']$ or $\rho_0 = \rho(t_0) = \max\limits_{t' \leqq t \leqq t''} \rho(t) > \delta$. Put $z(t') = z'$, $z(t'') = z''$, $z(t_0) = z_0$. If

$$z_0 z' = \min\big(h(z', z_0), h(z_0, z'')\big) \leqq \alpha \quad \text{then} \quad \rho_0 < \rho(t') + 2\alpha\beta = \delta + \alpha\beta.$$

If the minimum exceeds α we apply (d) and find

$$\beta^2 \geqq \frac{h(z', z_0) + h(z_0, z'')}{h(z', z'')}.$$

Then $\delta + 2h(z', z_0) \geqq \rho_0$, hence $2h(z', z_0) \geqq \rho_0 - \delta$, $2h(z_0, z'') \geqq \rho_0 - \delta$ and

$$\beta^2 \geqq \frac{\rho_0 - \delta}{|\zeta(t') - \zeta(t'')| + 2\delta} = \frac{\rho_0 - \delta}{4\delta},$$

so $\rho_0 \leqq 4\delta\beta^2 + \delta$ and thus in both cases

$$\rho_0 < \delta(4\beta^2 + 1) + \alpha\beta = \gamma'.$$

Since we know that a sequence $t_k \to \infty$ with $\rho(t_k) \leqq \delta$ exists this proves $\rho(t) < \gamma'$ for all $t \geqq 0$. The same γ' works for receding rays.

The remainder is simple. First we show that the initial point p_∞ of a line L cannot coincide with its endpoint. If this were the case take any point a on L. It decomposes L into a receding ray $R(p_\infty, a)$ and a progressing ray $R(a, p_\infty)$ and we could conclude that

$$L \subset U_h(R_h(a, p_\infty), \gamma').$$

Then sequences $b_\nu \in R(p_\infty, a)$ with $b_\nu a \to \infty$ and $c_\nu \in R(a, p_\infty)$ with $a c_\nu \to \infty$ and bounded $h(b_\nu, c_\nu)$ would exist which contradicts (a), because $b_\nu a + a c_\nu = b_\nu c_\nu$.

To prove the existence of $L(p_\infty, q_\infty)$ for $p_\infty \neq q_\infty$ take on $L_h(p_\infty, q_\infty)$ a sequence $a_\nu \to p_\infty$. Then any $R(a_\nu, q_\infty)$ satisfies

$$R(a_\nu, q_\infty) \subset U_h(R_h(a_\nu, q_\infty), \gamma') \subset U_h(L_h(p_\infty, q_\infty), \gamma').$$

This yields readily that a subsequence of $\{R(a_\nu, q_\infty)\}$ converges to a line $L(p_\infty, q_\infty)$.

Finally observe that for any a, b in H^n

$$U_h(R_h(a, q_\infty), \gamma' + h(a, b)) \supset U_h(R_h(b, q_\infty), \gamma')$$

and hence

(i) $$U_h(R_h(a, q_\infty), \gamma' + h(a, b)) \supset R(b, q_\infty) \cup R(q_\infty, b).$$

Consider $L_h = L_h(p_\infty, q_\infty)$ and any $L = L(p_\infty, q_\infty)$. Let $a \in L$ and $b_\nu \in L$ with $b_\nu \to q_\infty$. If c is any point of L_h then (i) shows that the subray $R(b_\nu, q_\infty)$ of L lies for large ν in $U_h(R_h(c, p_\infty), \gamma' + \delta_\nu)$, where δ_ν is the hyperbolic distance of b_ν from $R_h(a, q_\infty)$. But $b_\nu \in U_h(R_h(a, q_\infty), \gamma')$ and $R_h(c, q_\infty)$ and $R_h(L, q_\infty)$ have distance 0, so

$$b_\nu \in U_h(L_h(p_\infty, q_\infty), \gamma' + 1) \qquad \text{for large } \nu$$

and (i) yields that $R(p_\infty, b_\nu) \cup R(b_\nu, q_\infty) = L$ lie in $U_h(L_h(p_\infty, q_\infty), \gamma' + 1)$. Thus Theorem 1 is satisfied with $\gamma = \gamma' + 1$.

The reader would probably like to know *the problem treated by Efremovič and Tihomirova* ⟨1⟩. They consider an *equimorphism* ϕ of a hyperbolic space H^n on itself, which means that for a given $\varepsilon > 0$ a $\delta(\varepsilon)$ exists such that $h(x, y) < \delta(\varepsilon)$ and $h(\phi x, \phi y) < \delta(\varepsilon)$ imply respectively $h(\phi x, \phi y) < \varepsilon$ and $h(x, y) < \varepsilon$. Our arguments show that it suffices that ϕ is a homeomorphism and that the hypothesis holds for one ε (our α). The main result is that ϕ can be extended to a homeomorphism of $H^n \cup K$ onto itself.

In the proof the following results are established: Each $\phi R_h(a, p_\infty)$ has an endpoint p_∞^ϕ, each $\phi L_h(p_\infty, q_\infty)$ has two distinct endpoints $p_\infty^\phi, q_\infty^\phi$ and there is a γ such that

$$U_h(R_h(\phi a, p_\infty^\phi), \gamma) \supset \phi R_h(a, p_\infty)$$

and

$$U_h(L_h(p_\infty^\phi, q_\infty^\phi), \gamma) \supset \phi L_h(p_\infty, q_\infty).$$

The authors are not concerned with the existence of $R_h(a, p_\infty)$ for a given ϕa and p_∞^ϕ, but this follows from our discussion. In their arguments $h(\phi x, \phi y)$ plays the role of our xy, so that the difference is this: they deal with the hyperbolic distance of two new points, whereas we consider a new distance of the old points.

We now come to the *principal application of Theorem* (1). A closed hyperbolic space form is a compact manifold which can be provided with a locally hyperbolic metric, or equivalently (see for example G (30.1)), whose universal covering space is a hyperbolic space, H^n. The metric can then be chosen with curvature -1.

For $n=2$ this comprises all compact surfaces except the sphere, the projective plane, the torus and the one-sided torus. Although for $n>2$ the spherical space forms have all been determined and much is known regarding the euclidean forms (see Wolf ⟨1⟩), very little is known about hyperbolic space forms, except that compact hyperbolic space forms exist, see Löbell ⟨1⟩.

Let R be such a form provided with a not necessarily symmetric intrinsic distance xy. We also introduce in R a locally hyperbolic metric $h(x, y)$ of curvature -1. Lifting both metrics in the universal covering space $\bar{R}$ of R we obtain a metric $\bar{x}\bar{y}$ derived from R and H^n with a hyperbolic metric $h(\bar{x}, \bar{y})$ derived from $h(x, y)$. The metrics $\bar{x}\bar{y}$ and $h(\bar{x}, \bar{y})$ satisfy the hypothesis of Theorem (1), in fact α' exists for a given $\alpha>0$. This follows at once from the existence of compact fundamental domains, see G (34.1). If we denote as progressing (receding) half geodesic in R a partial geodesic $x(t)$ $(t \in M)$ with $M=\{t \geqq 0\}$ $(M=\{t \leqq 0\})$ then we deduce from Theorem (1):

(2) *Theorem. If a progressing (receding) half geodesic in R lifts into a progressing (receding) ray then this ray has an end (initial) point on K. If the geodesic in R lifts into a straight line in $\bar{R}$ then this line has an initial and an endpoint and these are distinct.*

For a given $a \in R$ and $\bar{a} \in \bar{R}$ over a and a given $p_\infty \in K$ there is a progressing (receding) half geodesic $x(t)$ in R with $x(0)=a$ which lifts into a $R(\bar{a}, p_\infty)$ $(R(p_\infty, \bar{a}))$. For given $p_\infty \neq q_\infty$ on K there is a geodesic in R which lifts into a line $L(p_\infty, q_\infty)$.

There is a $\gamma>0$ such that for any $\bar{a} \in \bar{R}$ and $p_\infty \neq q_\infty$

$$R(\bar{a}, p_\infty) \cup R(p_\infty, \bar{a}) \subset U_h(R_h(\bar{a}, p_\infty), \gamma)$$

$$L(p_\infty, q_\infty) \subset U_h(L_h(p_\infty, q_\infty, \gamma)).$$

The existence of so many half geodesics and geodesics in R which behave uniformly like hyperbolic half geodesics and geodesics is most surprising, in particular because we did not assume U or P.

We may consider the E^n *as an open hemisphere with boundary* K *in the n-dimensional spherical space* S^n. The lines in E^n then appear as semi great circles with antipodal endpoints on K. Assume that E^n carries besides the euclidean metric $e(x, y)$ an intrinsic metric xy for which constants α, α' as in Theorem (1) exist, then a ray with respect to xy does not necessarily have an endpoint, as observed by Efremovič and Tihomirova ⟨1⟩. Thus their method cannot be applied to closed euclidean space forms, in fact, it rests entirely on the strong divergence of intersecting hyperbolic lines.

On the other hand, Hedlund [1] showed that *Morse's method can be extended to the 2-dimensional torus* T with a Riemannian metric and this was generalized by Zaustinsky ⟨2⟩ to intrinsic metrics on T satisfying P and U. These conditions as well as $\dim T=2$ are used in an essential way. One obtains a theorem analogous to (2) with the obvious appropriate changes.

If xy is an intrinsic metric on T, we provide T also with a locally euclidean metric $e(x, y)$, lift both metrics into E^2 realized as subset of S^2, obtaining $\bar{x}\bar{y}$ and $e(\bar{x}, \bar{y})$. Then, for example, if a geodesic in T lifts into a straight line with respect to $\bar{x}\bar{y}$, this line has an initial and an endpoint which are antipodes. For any pair p'_∞, p_∞ of antipodes on K there is a geodesic in T which lifts into a $L(p'_\infty, p_\infty)$. There is a constant γ such that for any $L(p'_\infty, p_\infty)$

$$L(p'_\infty, p_\infty) \subset U_e\big(L_e(p'_\infty, p_\infty), \gamma\big).$$

These facts also hold for the one-sided torus because it has a torus as two-sheeted covering space.

13. Axes of Motions and Closed Geodesics

In the introduction to this chapter it is explained why axes of motions are interesting.

Let ϕ be a motion of the G-space R. If a straight line L exists, which ϕ maps on itself without fixed points then ϕ_L (the restriction of ϕ to L) has the form $x(t) \rightarrow x(t+\alpha)$ $(\alpha \neq 0)$ if $x(t)$ represents L. The orientation of L in which $x(t+\alpha)$ follows $x(t)$ is the *oriented axis* L^+ *of* ϕ, and we will always choose the representation $x(t)$ such that $\alpha > 0$. The opposite orientation of L is an axis of ϕ^{-1}. Also, L^+ is an axis of ϕ^ν for $\nu > 0$[3].

[3] The results (1) to (8) which follow are taken from Busemann and Pedersen ⟨1⟩.

(1) *If x lies on an axis of ϕ then*

$$x\phi x=\inf_{y\in R} y\phi y, \quad \textit{generally } x\phi^\nu x=\inf_{y\in R} y\phi^\nu y \textit{ for } \nu\neq 0.$$

Therefore ϕ^ν has no fixed point.

For $\nu=1$ this follows for $x\in L$, $y\in R$ from

$$x\phi^k x=k\,x\phi x\leqq xy+\sum_{\mu=1}^{k}\phi^{\mu-1}y\phi^\mu y+\phi^k y\phi^k x=2xy+k\,y\phi y$$

after division by k and letting $k\to\infty$. The general case $\nu\neq 0$ is a corollary.

For terminology and results used in the following see G, Section 22 and p. 138, for the definition of co-ray also Section 17, part c, here. Two oriented lines H^+ and L^+ in a G-space are parallel if each positive subray of $H^+(L^+)$ is a co-ray to a positive subray of $L^+(H^+)$ and also each negative subray of $H^+(L^+)$ is a co-ray to a negative subray of $L^+(H^+)$.

(2) *If the motion ϕ has an axis H^+ and maps the straight line L on itself, then L is an axis of ϕ. If L^+ is the oriented axis then H^+ and L^+ are parallel.*

By (1) ϕ has no fixed point, so L is an axis.

Choose a point q on L^+ and a positive subray S of H^+. Put $\phi^\nu q=q^\nu$, $\phi^\nu p=p^\nu$. For a limit sphere $K_\infty(q,S)$ we have $\phi^\nu K_\infty(q,S)=K_\infty(q^\nu,S)$, because $K_\infty(q^\nu,\phi^\nu S)=K_\infty(q^\nu,S)$. Moreover, $K_\infty(q,S)$ intersects H^+ in a point p (this is contained in G (22.14)). By (1) for $\nu>0$

$$p^{-1}p^\nu=q^{-1}q^\nu\geqq q^{-1}K_\infty(q^\nu,S)\geqq p^{-1}p^\nu,$$

so that q^ν is a foot of q^{-1} on $K_\infty(q^\nu,S)$. Therefore q^ν is for $\nu>1$ the only foot of q on $K_\infty(q^\nu,S)$, see G (20.6), and thus the co-ray to H^+ from q is the positive subray of L^+ beginning at q.

In general, H^+ will not be an axis of ϕ if it is an axis of ϕ^ν for some $\nu>1$. But in an important special case this is correct:

(3) *Let ϕ be an orientation preserving motion of a G-plane. If H^+ is an axis of ϕ^ν ($\nu>1$), then H^+ is also an axis of ϕ.*

Put $H_1^+=\phi H^+$ then

$$\phi^\nu H_1^+=\phi^\nu\phi H^+=\phi\phi^\nu H^+=\phi H^+=H_1^+,$$

so H_1^+ is an axis of ϕ^ν and hence by (2) parallel to H^+. This implies that either $H_1^+=H^+$ and the assertion or $H_1\neq H$. The latter is impossible. For, if Q is the half plane bounded by H and containing H_1 then, since ϕ preserves the orientation and $H_1^+=\phi H^+$, it maps Q on the half plane ϕQ

bounded by H_1 and not containing H. Repetition of this argument would yield that $\phi^\nu Q$ is a half plane not containing $H, \phi H, \ldots, \phi^{\nu-1} H$, contradicting $\phi^\nu H = H$.

In contrast to straight spaces (see G (32.3)) $0 < x\phi x = \inf_{y \in R} y\phi y$ *does not*, in general, imply that the points $\phi^\nu x$ lie on an axis. But a part of this assertion is true

(4) *For the motion* ϕ *let a point* p *with*

$$0 < p\phi p = \inf_{x \in R} x\phi x$$

exist. Then for any segment T *from* p *to* ϕp *the curve* $\bigcup_{-\infty}^{\infty} \phi^\nu T$ *is a geodesic* ζ *with* $\phi\zeta = \zeta$ *and* $y\phi y = p\phi p$ *for* $y \in \zeta$

Let y be an interior point of T. Then

$$p\phi p = py + y\phi p = y\phi p + \phi p\phi y \geqq y\phi y \geqq p\phi p,$$

hence $(y\phi p\phi y)$, so that $T' = T(y, \phi p) \cup T(\phi p, \phi p)$ is a segment and $\bigcup \phi^\nu T$ is a geodesic ζ. That $\phi\zeta = \zeta$ is obvious.

In one important case we can conclude that ζ is a straight line:

(5) *Theorem. If* ϕ *is an orientation preserving motion of a G-plane,* p *a point for which*

$$0 < p\phi p = \inf_{x \in R} x\phi x,$$

then the segment $T = T(p, \phi p)$ *is unique and* $\zeta = \bigcup_{-\infty}^{\infty} \phi^\nu T$ *is a straight line, and hence an axis of* ϕ.

The proof is longer than the preceding ones. Put $p^\nu = \phi^\nu p$, $T^\nu = T\phi^\nu T$ etc.

(a) *T is unique.* If two segments S, T from $p^0 = p$ to p^1 existed then by (4) both $\bigcup S^\nu$ and $\bigcup T^\nu$ would be geodesics, hence would cross each other at p^1. Therefore T traversed from p to p^1 followed by S traversed from p^1 to p would be an oriented simple closed curve whose image under ϕ is a curve with the opposite orientation.

(b) *G has no multiple points.*

Otherwise G would contain a simple monogon M. Because of (4) we may assume that p is the vertex of M. There are two possibilities:

1) $p = p^\nu$ for some $\nu \geqq 2$. Then $T^{\nu-1} = T(p^{\nu-1}, p^\nu)$, $T^\nu = T(p^\nu, p^{\nu+1}) = T(p, p^1) = T$ and $M = \bigcup_{=0}^{\nu-1} T$ or G is a simple closed geodesic and ϕ maps M and its interior on itself. Then ϕ would have a fixed point.

2) p is not a p^ν ($\nu \geqq 2$). Then $p^{\nu-1}p^0p^\nu$ for some ν. The proof of (4) then yields $(p\,p^\nu p^1)$ and by (a) the only segment from p to p^1 is $T(p, p^\nu) \cup T(p^\nu, p^1)$, which coincides with T. So ζ is again a simple closed geodesic, which leads to the same contradiction.

(c) *The arc* A^ν *of* ζ *from* p *to* p^ν ($\nu \geqq 1$) *is a segment.* This is correct for $\nu=1$. Assume $A^{\nu-1}$ to be a segment. If A^ν were not a segment, draw a segment S from p to p^ν. Then $p\,p^\nu = \text{length } S < \nu\, p\, p^1$. The intersection of S and A^ν consists of the points p and p^ν only because $A^{\nu-1}$ is a segment and $A^\nu = A^{\nu-1} \cup T(p^{\nu-1}, p^\nu)$.

Since G has no multiple points and ϕ preserves the orientation S and $S^1 = \phi S$ intersect at a point q. With $q^1 = S^1 \cap S^2$ and $p^1 p^\nu = (\nu-1)\,p\,p^1$ we would find

$$\nu\, p\, p^1 > p^1 p^{\nu+1} = p^1 q + q\, q^1 + q^1 p^{\nu+1} = p^1 q + q\, q^1 + q\, p^\nu$$
$$\geqq p^1 p^\nu + q\, q^1 = (\nu-1)\, p\, p^1 + q\, q^1$$

or $p\,p^1 > q\,q^1$ contradicting the hypothesis.

This relatively simple theorem allows us to generalize various results which were rather laboriously proved by others. Consider an orientable G-surface R which is not simply connected, i.e., topologically, any orientable surface other than the sphere or the plane. The universal covering space $\bar{R}$ of R is a G-plane and we interpret the fundamental group Δ of R as the group of motions of $\bar{R}$ lying over the identity of R. Because R is orientable the motions in Δ preserve the orientation.

Using the relations between free homotopy and closed geodesics discussed in G, Section 32, we find that, if in the (nontrivial) free homotopy class $\{K\}$ given by the closed curve K a shortest curve exists, then it is a closed geodesic G, and by (5) the curves over G are straight lines, which are axes of suitable elements of Δ in the class of conjugate elements determined by $\{K\}$.

Because an axis of ϕ is also one of ϕ^ν the curve G^ν (i.e., G traversed ν times) is a shortest curve in $\{K^\nu\}$ and by (3) a shortest curve in $\{K^\nu\}$ is a shortest curve in $\{K\}$ traversed ν times. Orientability is essential: the last assertion is false for a Moebius strip with a euclidean metric. We have proved:

(6) *Theorem. Let R be an orientable G-surface which is not simply connected. If the nontrivial free homotopy class $\{K\}$ in R contains a shortest curve G then the curves over G in $\bar{R}$ are straight lines and axes of motions of the fundamental group Δ.*

Moreover, G^ν is a shortest curve in $\{K^\nu\}$ $(\nu \neq 0)$ and a shortest curve in $\{K^\nu\}$ has the form G^ν where G is a shortest curve in $\{K\}$.

If R is compact then every nontrivial $\{K\}$ contains a shortest curve and every motion $(\neq 1)$ in Δ has an axis.

This theorem was proved for compact orientable surfaces with Riemannian metrics by Morse [1] for genus greater than 1 and by Hedlund [1] for the torus. (6) and Section 12 contain the essential parts of the two mentioned papers and thus solve Problem (32) in G, p. 406.

The present methods often prove very efficient in determining the distribution of conjugate points[4] *in given cases.* As an example we take a *G-torus* T with a one-parameter group of motions $\{\psi_t\}$ $(-\infty<t<\infty$, $\psi_{t_1}\psi_{t_2}=\psi_{t_1+t_2})$. This contains as special case the intrinsic metric on a surface in E^3 obtained by revolving a closed plane Jordan curve K about a line L which lies in the plane of K but does not intersect K, provided the metric satisfies the conditions for a G-space. A sufficient, but by no means necessary, condition is that K be of class C^2.

In the case where K is a circle the geodesics were given explicitly by Bliss ⟨1⟩. His interesting results on conjugate points suggested the question whether the behaviour of the geodesics remains qualitatively the same when the circle is replaced by another curve. The problem was taken up by Kimball ⟨1⟩. To make the calculations manageable he had to assume that K is convex analytic and symmetric with respect to a plane normal to L. Our method works for any K (of class C^2) and with much less effort.

To elucidate the discussion which follows we observe that the points of K with minimal distance from L yield closed geodesics of minimal length in their homotopy class. They are orbits under rotation about L and are, by (5), lifted into straight lines in the universal covering space.

We return to the general case of a G-torus T with a group $\{\psi_t\}$ of motions. If the metric of T is Minkowskian then T posses one-parameter groups of motions whose orbits are geodesics which are not closed. The converse holds:

(7) *If a G-torus posses a one-parameter group of motions and no orbit of this group is a closed geodesic, then the metric is Minkowskian.*

This and the following statements are rather easily derived from (1) to (5), but we refer for proofs to Busemann and Pedersen ⟨1⟩.

If $\{\psi_t\}$ has an orbit which is a geodesic, then the universal covering space is a G-plane R with a group of motions $\{\phi_t\}$ lying over $\{\psi_t\}$, and $\{\phi_t\}$ has orbits which are straight lines. The lines are parallel and form a closed set in R. *The distribution of conjugate points is then the following, in precise analogy to the results of Bliss for the ordinary torus:*

A geodesic in R which intersects one of the straight orbits is a straight line and hence intersects all straight orbits, consequently all orbits of $\{\phi_t\}$.

[4] For the relations of G-spaces to the calculus of variations and its terminology see G p.p. 113–116. In particular, straightness of the universal covering space of R is equivalent to the absence of conjugate points in R.

A point p which does not lie on a straight orbit lies in a smallest strip bounded by two such orbits A^+, B^+. Through p there are two straight lines H^+, L^+ such that $H^+(L^+)$ is an asymptote to $A^+(B^+)$ and $H^-(L^-)$ is an asymptote to $B^-(A^-)$ [5].

H and L define two angular domains D_1, D_2, where D_1 contains A and B. Each geodesic through q which in a neighborhood of q lies in D_i stays in D_i $(i=1,2)$. If it lies in D_1 it is a straight line and intersects all orbits. If in D_2 it has no multiple point at p, hence no multiple point anywhere and no half geodesic on it is a ray. This means in the language of the calculus of variation is, that every point of the geodesic is followed and preceded by conjugate points.

This discussion and (7) contain

(8) *If a G-torus possesses a one-parameter group of motions* $\{\psi_t\}$ *and all orbits of* $\{\psi_t\}$ *are geodesics then the universal covering space is straight* (but, in contrast to the Riemannian case, not necessarily Minkowskian, see G (33.5)).

14. Plane Inverse Problems. Higher Dimensional Collineation Groups

The question when for a two-parameter system of curves in the plane a Riemannian metric exists locally for which the given curves are the geodesics has been open for a long time and is still unsolved, see Blaschke and Bol ⟨1, Section 29⟩. An answer will probably state, at least implicitly, that the curves must satisfy Desargues's Theorem infinitesimally to a higher degree than arbitrary smooth curve systems.

That these can, locally, be regarded as the extremals of a variational problem was shown by Darboux ⟨1, Section 605⟩ as follows: He considers a curve system in the (x, y)-plane as given by a differential equation $y''=\phi(x, y, y')$. The Euler Equation for $f(x, y, y')$ is

$$f_{y'y'}\, y''+f_{yy'}\, y'+f_{yx}-f_y=0.$$

Replacing y'' by $\phi(x, y, y')$ this yields a partial differential equation for $f(x, y, y')$ with x, y, y' as independent variables. The given curves are extremals for f, i.e., solve the equation.

This method has many shortcomings: our (and Darboux's) situation demands the parametric form, a positive f and extremals which yield strict minima, i.e., at least the Legendre Condition. Even if all these conditions are satisfied the solution will only be local. We show here that an idea from *integral geometry* provides very simple solutions satisfying all requirements.

[5] H^- is the opposite orientation of H^+. In all 4 cases this is true in the strong sense that, for example, $xA\to 0$ if x traverses H^+ in the positive direction, so that A^+ is also an asymptote to H^+, see G (22.22).

On a G-surface each point has a convex neighborhood C homeomorphic to E^2, for example the interior of a small geodesic triangle. If we denote the open segments in C with endpoints on the boundary of C as lines, then these form a curve system Σ with the following topological properties. Σ is defined in $\tilde{E}$ homeomorphic to E^2. Each curve in Σ (called line) has the form $p(t)$ $(-\infty<t<\infty)$ with $p(t_1)\neq p(t_2)$ for $t_1\neq t_2$ and no sequence $p(t_i)$ with $|t_i|\to\infty$ converges. Through two distinct points in $\tilde{E}$ there is exactly one line.

With their "natural" topology (which is equivalent to the Hausdorff closed limit, see G, Section 11) *the lines in Σ form a two-dimensional manifold*. For, if $L\in\Sigma$ and a, b are distinct points of L, let I_a, I_b be open intervals on lines through a and b which contain a and b respectively and do not intersect ($I_a\cap L=a$, $I_b\cap L=b$, $I_a\cap I_b=\emptyset$). The lines intersecting both I_a and I_b form evidently a set homeomorphic to $\tilde{E}$ and a neighborhood of L.

It is not hard to prove (see Busemann and Salzmann ⟨1, (6)⟩) that Σ is topologically the punctured projective plane or the (open) Moebius strip, but this information will not be used. All we need is *the existence of a measure in Σ for which a nonempty open set has positive measure, the lines intersecting a compact pointset form a set of finite measure, which vanishes when the set consists of a single point.*

If μ is any such measure and μA the measure of the set A of lines in Σ we define for any two distinct points a, b as $T(a, b)$ the arc of the line $L(a, b)$ through a and b with endpoints a, b and a *distance* by $aa=0$ and

$$ab=\mu\{L|L\cap T(a, b)\neq\emptyset\}.$$

Because of $\mu\{L|b\in L\}=0$ it is clear that

$$ab+bc=ac \qquad \text{if } b\in T(a, c).$$

If b does not lie on $T(a, c)$ then each line intersecting $T(a, c)$ intersects either $T(a, b)$ or $T(b, c)$ or both, so $ab+bc\geqq ac$. Since there is an open set of lines intersecting $T(a, b)\cup T(b, c)$ which do not intersect $T(a, c)$ we have

$$ab+bc>ac \qquad \text{if } b\notin T(a, c).$$

Thus ab defines a metric space and the $T(a, b)$ are segments and the only ones for this distance. For the local problem we choose μ such that $\mu\Sigma$ is finite and thus obtain in a few lines *a solution in the large, i.e., in the entire given convex neighborhood.* Observe that for smooth Σ in the (x, y)-plane well behaved measures trivially exist and that from these we obtain by differentiation smooth integrands $F(x, y, \dot{x}, \dot{y})$ for which the given lines are the minimizing curves.

If we are interested in the global problem, where we want to make Σ the geodesics of a G-plane, finite compactness becomes an additional problem which is frequently as difficult as finding a metric for which the given curves minimize length without being isometric to the entire real axis. In the present case it can be solved by combining different measures, see Salzmann ⟨1, p. 13⟩. The problem was solved in G, Section 11 by a different method consisting in a superposition of quasimetrics. The deeper problems to be considered later involve both methods.

First we use the integral geometric method to solve in a surprisingly simple way a problem which is answered by (1) below, but which was left open in G. It was first solved by Skornyakov ⟨1⟩ by generalizing the superposition of quasimetrics, which becomes quite involved, then somewhat simpler, although still in substantially the same way, by the method used in the proof of (7.1).

(1) *Theorem. Let Σ be a system of closed Jordan curves in the projective plane $\tilde{P}$ such that two distinct points of $\tilde{P}$ lie on exactly one curve of Σ. Then $\tilde{P}$ can be metrized as a G-space with the curves in Σ as geodesics.*

This time Σ is a compact manifold and in fact homeomorphic to $\tilde{P}$, but as before, this information is not needed. We use any measure μ with the same properties as above and define for any subarc A of a curve in Σ

$$\mu A = \mu\{L | L \cap A \neq \emptyset\}.$$

Since all curves in Σ except L_0 intersect L_0 exactly once we have

$$\mu L_0 = \mu\{L | L \cap L_0 \neq \emptyset\} = \mu \Sigma.$$

We define $aa = 0$ and ab for $a \neq b$ as the minimum of the lengths of the two arcs with endpoints a, b on the curve in Σ through a and b. Then the same arguments as above show that locally the arcs of curves in Σ are the only segments. Therefore the axioms of a G-space are satisfied and the curves in Σ are the geodesics.

The entirely different character of the corresponding Riemannian problem is evident from the fact, contained in Green ⟨1⟩, that the elliptic plane is the only Riemann space satisfying the hypothesis of (1).

If a curve system Σ either in $\tilde{E}$ or in $\tilde{P}$ is given as above, then a topological map of $\tilde{E}$ or $\tilde{P}$ which takes Σ into itself is called a Σ-collineation.

A group Γ of Σ-collineations is denoted by $\{\Gamma, \Sigma\}$ and is called metric, if a metrization of $\tilde{E}$ or $\tilde{P}$ as a G-space exists for which the curves in Σ are the geodesics and Γ is a group of motions. Since the group of all motions of a G-space is locally compact and is compact when the space is compact, we can and will assume that Γ is in the case of $\tilde{E}$ locally compact and compact in the case of $\tilde{P}$.

The determination of the metric $\{\Gamma, \Sigma\}$ with $\dim \Gamma \geqq 2$ is reduced to results in G by some simple lemmas.

(2) *If an n-dimensional manifold carries a metric xy and a group of motions Γ with an n-dimensional orbit then Γ is transitive.*

For, if the orbit Γ_p has dimension n, then it contains interior points, see Hurewicz and Wallman [1, p. 40], so $S(q,\rho)\subset\Gamma_p$ for a suitable $q\in\Gamma_p$ and $\rho>0$. Therefore $S(x,\rho)\subset\Gamma_p$ for each $x\in\Gamma_p$, so that Γ_p is both open and closed.

A rotation of a G-surface R about a point p is a motion which leaves p fixed and induces for $0<\rho<\rho(p)$ an orientation preserving map of $S(p,\rho)$ on itself. This formulation is necessary because $S(p,\rho)$ is homeomorphic to a disk and orientable, whereas R may not be orientable. We say that all rotations about p exist if for $xp=yp$ a rotation about p exists which takes x into y. It suffices, of course, that this be true if $xp=yp<\rho(p)$. Trivially:

(3) If the G-surface R possesses all rotations about p and ϕ is a motion then all rotations about ϕp exist.

(4) *Let the G-surface R possess all rotations about p and about q where* $0<pq<\sup_{x\in R} px$. *Then R is elementary or elliptic.*

For, if $\rho=pq$ then $K(p,\rho)$ separates R and hence contains a continuum. The rotations about p are transitive on $K(p,\rho)$, and since $q\in K(p,\rho)$ all rotations about any $x\in K(p,\rho)$ exist. There are x, y on $K(p,\rho)$ with $0<xy<\rho(x)/2$, so all rotations about $y\in K(x,xy)$ exist, whence it follows that the group of all motions of R has a two-dimensional orbit. By (2) the group is transitive and by (3) all rotations about any point of R exist, so that R is elementary or elliptic.

(5) *Theorem. Let Γ be the group of all motions of a G-surface R. If* $\dim\Gamma=3$ *then R is euclidean, hyperbolic, spherical or elliptic. If* $\dim\Gamma=2$ *then R is either the plane with a Minkowskian but not euclidean, or a quasihyperbolic but not hyperbolic metric, or a cylinder or torus with a Minkowskian, possibly euclidean, metric.*

For, if $\Gamma(p)$ is the stability group of p then

$$\dim\Gamma=\dim\Gamma_p+\dim\Gamma(p).$$

Obviously $\dim\Gamma(p)\leqq 1$ and $\dim\Gamma_p\leqq 2$. Therefore $\dim\Gamma=3$ implies $\dim\Gamma_p=2$, $\dim\Gamma(p)=1$, so that Γ is transitive on R and all rotations about every point exist. R is elementary or elliptic.

If $\dim\Gamma=2$ then $\dim\Gamma(p)=1$ is impossible, because $\dim\Gamma_p=1$ and (4) would show that R is elementary or elliptic and hence $\dim\Gamma=3$. Thus $\Gamma(p)$ is discrete and Γ is transitive (its identity component is simply transitive). The classification of all G-surfaces with transitive groups of motions in G (52.7) yields the second part (5).

This theorem implies for the metric $\{\Gamma, \Sigma\}$ in $\tilde{E}$ or $\tilde{P}$ that $\dim \Gamma = 2$ is impossible for P^2 [6], and for all $\{\Gamma, \Sigma\}$ which are metric with $\dim \Gamma = 2$ it furnishes not only the curve systems but also the metrics which are either Minkowskian or quasihyperbolic. There are as many nonisometric plane Minkowski metrics as there are affinely nonequivalent strictly convex curves with a center, see G (17.10). The precise freedom of choice in the case of the quasihyperbolic metric is discussed in Section 16.

If a collineation group $\{\Gamma, \Sigma\}$ is metric and $\dim \Gamma \leqq 1$, then for given Σ the choice of metrics invariant under Γ is so large, that *the determination of all metrics ceases to be a reasonable question, and we only determine the topological structure of Σ in relation to Γ.* In this sense we will find all one-dimensional $\{\Gamma, \Sigma\}$ and the most interesting discrete $\{\Gamma, \Sigma\}$.

One may pose the problem of finding all curve systems in $\tilde{E}$ or $\tilde{P}$ which possess (not necessarily metric) collineation groups of a certain dimension. This has been done for Σ in $\tilde{E}$ which satisfy the parallel axiom, principally by Salzmann, see his report ⟨1⟩. It turns out that the question is reasonable for $\dim \Gamma \geqq 3$, but not for $\dim \Gamma < 3$ because there are too many possibilities.

15. One-Dimensional and Discrete Collineation Groups

Turning to one-dimensional or discrete metric collineation groups we first prove a theorem on general G-spaces which is interesting in itself and which will allow us to settle the compact case at once and to reduce certain cases to others [7].

(1) *Theorem. Let Γ be a locally compact group of motions of the G-space R which is a normal subgroup of the group Γ^* of topological geodesic preserving maps of R on itself with compact Γ^*/Γ. Then R can be remetrized as a G-space R^* which has the same geodesics as R and for which the elements of Γ^* are motions.*

Let $\alpha = \phi\Gamma = \Gamma\phi \in \Gamma^*/\Gamma$ and put

$$\delta_\alpha(x, y) = \phi x\, \phi y.$$

This number depends only on α and not on the choice of ϕ in α because $\phi' \in \alpha$ implies $\phi' = \gamma\phi$ with $\gamma \in \Gamma$ and

$$\phi' x\, \phi' y = \gamma\phi x\, \gamma\phi y = \phi x\, \phi y.$$

[6] This also follows from a theorem of Montgomery and Zippin ⟨1⟩, which states that no proper subgroup of the group of rotations of E^3 about a point is transitive on a sphere about this point.

[7] The results for which no other reference is given are taken from Busemann and Salzmann ⟨1⟩ which will be quoted as *BS*.

Also, for $\psi \in \Gamma^*$ and $\beta = \psi \Gamma$

$$\delta_\alpha(\psi x, \psi y) = \psi \phi x \, \psi \phi y = \delta_{\beta\alpha}(x, y).$$

Using the measure or invariant integration on the compact group Γ^*/Γ (see Pontrjagin [1, Section 25]) we form

$$\delta^*(x, y) = \int \delta_\alpha(x, y)\, d\alpha,$$

which exists because $\delta_\alpha(x, y)$ depends continuously on x, y and α. It is clear that $\delta^*(x, y)$ satisfies the axioms for a metric space and is invariant under Γ^*, because for any $\psi \in \Gamma^*$

$$\delta^*(\psi x \, \psi y) = \int \delta_\alpha(\psi x \, \psi y)\, d\alpha = \int \delta_{\beta\alpha}(x, y)\, d\alpha = \delta^*(x, y).$$

Since Γ is locally compact and Γ^*/Γ is compact, there is a compact set $K \subset \Gamma^*$ such that $\Gamma^* = K \cdot \Gamma = \Gamma \cdot K$. If $\{x_\nu\}$ is a sequence of points in R for which $\delta^*(x_1, x_\nu) \leqq k$ for all ν, then for each ν a $\phi_\nu \in K$ with $\phi_\nu x_1 \, \phi_\nu x_\nu \leqq k$. Because $K x_1$ is compact, $\{\phi_\nu x_\nu\}$ is bounded in the original metric xy and therefore lies in a compact set C of R, so $\{x_\nu\}$ lies in the compact set $K^{-1} C$. Thus R^*, i.e., R with the metric δ^*, is finitely compact.

Let $x(t)$ represent a geodesic in R. For any t_0 we have

$$\rho_0 = \min_{\psi \ K} \rho(\psi\, x(t_0)) > 0.$$

If $t_1 < t_2 < t_3$ and $|t_i - t_0| \leqq \rho_0$ then with $x_i = x(t_i)$

$$\psi x_1 \, \phi x_2 + \psi x_2 \, \psi x_3 = \psi x_1 \, \phi x_3 \qquad \text{for } \psi \in K$$

because $\psi x(t)$ is, by hypothesis, a geodesic. Therefore $x(t)$ $(|t - t_0| \leqq \rho_0)$ is a segment, first for each metric δ_ψ with $\psi \in K$, then by $\Gamma^* = K\Gamma$ for all $\alpha \in \Gamma^*/\Gamma$ and finally for δ^*. Thus $x(t)$ is also a geodesic for δ^*, although t need not be arc length.

Applying this to the case where Γ consists of the identity only we find:

(2) *Each compact, and hence each finite, geodesic preserving group of topological maps of a G-space R on itself is a group of motions for a suitable remetrization of R as a G-space with the same geodesics.*

Since any system Σ in $\tilde{E}$ or $\tilde{P}$ forms the geodesics of a metrization of $\tilde{E}$ or $\tilde{P}$ as a G-space, we conclude from (2)

(3) *Theorem. Any compact collineation group $\{\Gamma, \Sigma\}$ in $\tilde{E}$ or $\tilde{P}$ is metric.*

This solves the problem for $\tilde{P}$ completely since, as previously observed, Γ may be assumed to be compact. Therefore we only consider $\tilde{E}$ and denote, as before, by $L(a, b)$ the line or curve in Σ through $a \neq b$ and by $T(a, b)$ the arc with endpoints a, b on $L(a, b)$.

The situation for noncompact Γ is quite involved and trying to find all metric $\{\Gamma, \Sigma\}$ is hopeless, at least at present. However, (1) yields some simplifications, for example:

(4) *If the subgroup $\{\Gamma', \Sigma\}$ of all orientation preserving collineations in $\{\Gamma, \Sigma\}$ is metric, then so is $\{\Gamma, \Sigma\}$.*

For, if Γ' is a proper subgroup of Γ then it has index 2 and hence is normal. Thus it suffices to consider orientation preserving collineations.

Consider a one-parameter group Γ_t of Σ-collineations $\psi_t(-\infty<t<\infty$, $\psi_{t_1}\psi_{t_2}=\psi_{t_1+t_2})$. For fixed α we denote the subgroup $\psi_{\nu\alpha}$ $(\nu=0, \pm 1, \ldots)$ by $\Gamma_{\nu\alpha}$.

From (1) we deduce further:

(5) *The one-parameter group $\{\Gamma_t, \Sigma\}$ is metric if, and only if, for some $\alpha \neq 0$ (and hence for each α) the group $\{\Gamma_{\nu\alpha}, \Sigma\}$ is metric.*

For, $\Gamma_t/\Gamma_{\nu\alpha}$ is compact and $\Gamma_{\nu\alpha}$ is a normal subgroup.

Next we observe

(6) *If the Σ-collineation ϕ has no fixed point then no ϕ^ν $(\nu \neq 0)$ does.*

Put $p^\nu=\phi^\nu p$. If p were a fixed point of ϕ^2 then ϕ would map $T(p, p^1)$ on itself interchaning p and p^1, hence have a fixed point on $T(p, p^1)$. We may assume that ϕ preserves the orientation, if not we operate with ϕ^2. If there were a first $m>2$ for which ϕ^m has a fixed point p, then ϕ would map the Σ-convex polygon $\bigcup_{i=1}^{m-1} T(p^i, p^{i+1})$ and its interior on itself, and therefore have a fixed point. (6) has the corollary:

(7) If $\{\Gamma_t, \Sigma\}$ is a one-parameter group of Σ-collineations then either all ϕ_t have the same fixed point or no ϕ_t $(t \neq 0)$ has a fixed point.

(3), (5) and (7) reduce the problem, when a one-parameter group is metric to the case of a cyclic group $\Gamma_\phi=\{\phi^\nu\}$, where ϕ is an orientation preserving Σ-collineation without fixed points.

Not every $\{\Gamma_\phi, \Sigma\}$ is metric. As one might have expected the concept of an axis again plays a decisive role. An oriented line $L^+ \in \Sigma$ is an axis of ϕ if $\phi L^+ = L^+$ and the orientation is such that ϕx follows x for εL^+. Let $L^+(x, y)$ denote the oriented line $L(x, y)$ with x preceding y. For any oriented line L^+ any any point p the lines $L^+(p, x)$ converge to a line A^+ where x traverses L^+ in the positive direction. A^+ is the asymptote to L^+ through p and is independent of p in the sense for $q \in A^+$ the limit of $L^+(q, x)$ is also A^+ (see G, Section 11). We prove:

(8) *If $\{\Gamma_\phi, \Sigma\}$ is metric and L^+ is an axis of ϕ then $L^+(p, \phi^\nu p)$ tends for any p and $\nu \to \infty$ to an asymptote to C^+.*

This implies, of course, that $L^+(p, \phi^{-\nu}p)$ tends for $\nu \to \infty$ to an asymptote to the opposite orientation of L^+.

Assume $p \notin L$. The line $L(p, \phi p)$ does not intersect L^+ since by G (32.12) the points $p^\nu = \phi^\nu p$ lie for $\nu > 1$ between $L(p, \phi p)$ and L, so $L(p, \phi^\nu p) \cap L = \emptyset$ for all $\nu \geqq 1$. If $r \in L$ then the intersection $L(p, p^\nu) \cap T(r, p_1)$ tends monotonically to a point q and $L^+(p, p^\nu) \to A^+ = L^+(p, q)$ with $A^+ \cap L^+ = \emptyset$. Since $L(p, p^\nu) \cap L(p', p^\nu) = p^\nu$ has no accumulation point, $A^+ \cap \phi A^+ = \emptyset$ and $\phi^{-1} A^+$ lies between L^+ and A^+. Therefore

$$B^+ = \lim_{\nu \to \infty} \phi^{-\nu} A^+$$

exists and is an axis of ϕ. If x traverses A^+ in the positive direction then $xB^+ \to 0$. By G (22.22) the lines A^+ and B^+ are asymptotes to each other and so are A^+ and L^+, see G (32.16).

We say that $\{\Gamma, \Sigma\}$ satisfies the *Asymptote Condition*, if for every orientation preserving $\phi \neq 1$ in Γ, every axis L^+ of ϕ and every point p the lines $L(p, \phi^\nu p)$ converge to an asymptote to L^+ when $\nu \to \infty$. We may reformulate (8) as

(9) *Theorem. If $\{\Gamma, \Sigma\}$ is metric, then it satisfies the asymptote condition.*

To see that a nonmetric $\{\Gamma, \Sigma\}$ need not satisfy the asymptote condition, we consider the system Σ constructed in G, Section 23 to exemplify asymmetry of asymptotes. Σ consists of the lines $x_2 = mx_1 + b$ with $m \leqq 0$ of the (x_1, x_2)-plane, the lines $x_1 = \text{const}$ and the translates of the branch in $x_1 < 0$ of the hyperbole $x_1 x_2 = -1$. It is invariant under all translations $\tau_a : x_i' = x_i + a_i$.

If $a_1 > 0$, $a_2 > 0$ then τ_a does not have axes. If $a_1 > 0$, $a_2 < 0$ then τ_a has the family of parallel lines $a_2 x_1 - a_1 x_2 = c$ as axes, so that Γ_{τ_a} satisfies in both cases the asymptote condition. But it evidently does not when $a_1 \neq 0$, $a_2 = 0$ or $a_1 = 0$, $a_2 \neq 0$.

Let a_1, a_2 be irrational numbers with $a_1 : a_2 \neq 1$. The translations $\tau_1 : x_i' = x_i + 1$ and τ_a generate a group Γ of translations of Σ which satisfies the asymptote condition since it does not contain translations parallel to the x_1- or x_2-axes, but Γ is not metric since τ_1 does not have an axis or Σ does not satisfy the parallel axiom, see G (33.1, 3). Thus the asymptote condition is not sufficient for $\{\Gamma, \Sigma\}$ to be metric. However, it is, in important special cases:

(10) *Theorem. If Γ_ϕ is the cyclic group generated by an orientation preserving Σ-collineation ϕ without fixed points, then $\{\Gamma_\phi, \Sigma\}$ is metric if (and only if) it satisfies the asymptote condition.*

Notice that $\{\Gamma_\phi, \Sigma\}$ is always metric when ϕ has no axis. The proof of (10) is, unfortunately, quite long and involved, therefore we refer the reader to BS, pp. 223–228. (5) and (10) imply

(11) *A one-parameter group* $\{\Gamma_t, \Sigma\}$ *of* Σ*-collineations without fixed points (for* $t \neq 0$*) is metric if and only if it (or* $\{\Gamma_{\nu\alpha}, \Sigma\}$ *for some* $\alpha \neq 0$*) satisfies the asymptote condition.*

This leads to a complete discussion of the one-dimensional collineation groups $\{\Gamma, \Sigma\}$ (see BS, pp. 228–234). The identity component is a one-parameter group Γ_t and $\dim \Gamma = 1$ implies that an element of Γ permutes the orbits of Γ_t. Using (1) twice one reduces the problem to the subgroup Γ^+ of those collineations in Γ which preserve the orientation of the plane and of the orbit space and finds if $\Gamma^+ \neq \Gamma_t$:

(12) *Theorem.* Γ^+ *is the extension of* Γ_t *by an infinite cyclic group* $\Gamma_\phi = \{\phi^\nu\}$*, i.e.,* $\Gamma^+ = \Gamma_\phi \Gamma_t$ *and* Γ^+ *is abelian or not according to whether* Γ_t *has an orbit (axis) which is a line in* Σ *or not.*

In the nonabelian case $\{\Gamma^+, \Sigma\}$ *is metric if (and only if) it or* $\{\Gamma_\phi, \Sigma\}$ *satisfies the asymptote condition.*

An abelian $\{\Gamma^+, \Sigma\}$ *is metric if and only if* Σ *satisfies the parallel axiom and for each* $\psi \in \Gamma^+$ $(\psi \neq 1)$ *the axes of* ψ *cover the plane.*

The fact that each Σ in $\tilde{E}$ is the system of geodesics for a G-space provides a solution of the inverse problem for metrics without conjugate points in the plane. The analogous question may be raised for *systems* Σ' *on a surface* S *which is not simply connected.* We pass to $\tilde{E}$ as the universal covering space of S and lift Σ' into a system Σ in $\tilde{E}$. Then Σ satisfies our conditions and is in addition invariant under the discrete group Γ_S of the Σ-collineations which correspond to the covering transformations of $\tilde{E}$ over the identity of S. So we have to answer the question: when is $\{\Gamma_S, \Sigma\}$ metric?

For the euclidean forms, i.e., the cylinder C, Moebius strip M, torus T, and one-sided torus or Klein bottle B, *we have complete solutions.* M and B have, respectively, C and T as two-sheeted covering spaces and $\Gamma_C(\Gamma_T)$ consists of the orientation preserving elements in $\Gamma_M(\Gamma_B)$. By (4) $\Gamma_M(\Gamma_B)$ is metric if (and only if) $\Gamma_C(\Gamma_T)$ is. (10) shows that Γ_C is metric if it satisfies the asymptote condition. The answer for Γ_T is given in G, Section 33: Σ must satisfy the parallel axiom and if a line in Σ contains p and ϕp $(\phi \in \Gamma_T)$ then it must contain all points $\phi^\nu p$ $(\nu = \pm 1, \pm 2, \ldots)$.

For the hyperbolic forms a complete answer is not available. Owing to the enormous number of possibilities, it seems hopeless to attempt finding one, unless there is a hitherto undiscovered ordering principle. Theorem (12) provides metrizations in very many special cases, also for noncompact surfaces, but a nearly, possibly entirely, complete solution is known only for the compact forms S. We realize the universal covering space of S as the *Klein model* of the hyperbolic plane H^2 with the curvature -1 in the interior of the unit circle K of the euclidean plane and denote the group of hyperbolic motions of H^2 by Γ_h. If S is metrized as a G-space without conjugate points, then the geodesics lift into a system of curves Σ in H^2 which is invariant under the subgroup Γ_S of Γ_h furnishing the covering transformations of H^2 over S (see G, Section 34) and Σ has the following three properties, see Sections 12, 13.

A. *Each curve in Σ or line is simple and two distinct points of H^2 lie on exactly one line.*

B. *A line has two distinct end points on K.*

C. *Any two distinct points on K are the endpoints of a line.*

The construction which follows is a *particularly nice application of integral geometry and comprises many curve systems on noncompact hyperbolic forms because it does not use the special structure of Γ_S, but only A, B, C.* Denote by β the map of Σ onto the set P of nonordered pairs of distinct points of K, obtained by mapping a line on its endpoints and by β_h the corresponding map for the hyperbolic lines. The kinematic measure for the hyperbolic lines induces a measure on the Borel subsets Q of P by

$$m(Q) = measure\ \beta_h^{-1}(Q).$$

For any Borel set M in H^2 put

$$P(M) = \beta\{L | L \in \Sigma,\ L \cap M \neq 0\}$$

and

$$\mu(M) = m(P(M)),\ xy = \mu(T(x,y)),\ xx = \mu(P(x)).$$

The set $P(M)$ is measurable because

α) *β is continuous. Hence β is an open map if it is bijective*, i.e., if the endpoints of a line determine the line uniquely.

β) $\mu(M) < \infty$ *for compact* M.

This follows from the easily established fact (see *BS*, p. 235) that in this case an $\eta(M) > 0$ exists such that the euclidean distance of the points in any pair in $P(M)$ exceeds $\eta(M)$.

γ) $xx = 0.\ 0 \leqq xy = yx < \infty.$

The second assertion follows from β. For a given $a \in K$ let $N(a)$ be the union of the rays from x to a. Then $N(a) \cap N(a') = \emptyset$ for $a \neq a'$ which proves $xx = 0$.

δ) $P(T(x,z)) \cup Q = P(T(x,y)) \cup P(T(y,z))$ with
$Q = Q(y; x, z) = P(T(x,y)) \cap P(T(x,z))$.

ε) $xy + yz \geqq xz$ *with inequality if, and only if,* $\mu(Q(y; x, z)) > 0$.

ζ) *Let $x \neq z$ and let no other line than $L(x,z)$ with the same endpoints as $L(x,z)$ exist, then*

$$xy + yz > xz \quad \textit{for any } y \notin L(x,z).$$

η) $xy + yz = xz$ *if* $y \in T(x,z)$.

For, if points a, b on K and lines L_1, L_2 exist, such that L_1 intersects $T(y, z)$, then there is also a line with endpoints a, b through y. So $Q(y; x, z) = P(y)$, and $\eta)$ follows from $\gamma)$ and $\varepsilon)$.

$\theta)$ *$\mu(R) = \infty$ for any ray in Σ.*

$\iota)$ *If Σ is invariant under a subgroup of Γ_h then so is xy.*

This follows from the invariance of the kinematic measure for the hyperbolic lines, which implies invariance of m.

From these observations we deduce two theorems:

(12a) *Theorem. Let Σ be a curve system in H^2 which satisfies A, B, C and is invariant under the subgroup Γ of Γ_h. If the map β is open, in particular, if the endpoints of a line determine the line uniquely, then $\{\Gamma, \Sigma\}$ is metric.*

For, $P(T(x, y))$ contains a nonempty open subset, so $xy > 0$ for $x \neq y$. Moreover, if x, y, z are noncollinear, then the set M of lines separating y from x and z is nonempty, open and contained in $Q(y, x, z)$, so $xy + yz > xz$ by $\varepsilon)$.

(13) *Theorem. Let Σ be a curve system in H^2 which satisfies A, B, C and is invariant under $\Gamma \subset \Gamma_h$. If $P(T(x, y))$ $(x \neq y)$ contains a nonempty open subset of P, then H^2 can be metrized by a distance invariant under Γ such that the curves in Σ are straight lines and $T(x, y)$ is the only shortest join of x and y whenever no line other than $L(x, y)$ with the same endpoints as $L(x, y)$ exists.*

We return to the special case $\Gamma_S \subset \Gamma_h$ of the covering transformations of a compact hyperbolic form. Then A, B, C hold. It is easy to construct examples, where axes of elements of Γ_S are not uniquely determined by their endpoints. Using an argument of Zaustinsky ⟨2⟩ one derives from the asymptote condition, that any line with the same endpoints as an axis is an axis. It seems probable but has not been proved, that any line not uniquely determined by its endpoints is an axis. Thus the following theorem may contain a complete answer for Γ_S, in any case it contains the most significant Σ.

(14) *Theorem. Let Γ_S be the group of covering transformations of a 2-dimensional compact hyperbolic space form and let Σ be a curve system in H^2 invariant under Γ_S which satisfies A. If $\{\Gamma_S, \Sigma\}$ is metric, then B, C and the asymptote condition hold. Conversely, if these conditions are satisfied and every line which is not an axis of an element of Γ_S is uniquely determined by its endpoints, then $\{\Gamma_S, \Sigma\}$ is metric.*

Since the pairs in P which are endpoints of axes are countable, it follows that $P(T(x, y))$ contains a nonempty open set. Therefore (13) is applicable. It is possible to add suitable quasimetrics invariant under

Γ_S to $\mu(T(x,y))$ such that the sum is a metric satisfying (14), see *BS*, pp. 238, 239. We mention, that according to Zauskinsky ⟨2⟩ the asymptote condition allows us to relax *B*; it implies that the endpoints of a line in Σ are distinct.

16. Bonnet Angles. Quasi-Hyperbolic Geometry

The Gauss-Bonnet Theorem is one of the principal tools for discussing questions in the large, in particular the behaviour of the geodesics, for two-dimensional Riemann spaces. Many of these results can be extended to *G*-surfaces (see *G*, Sections 42–44) using general angular measures instead of the Gauss-Bonnet Theorem.

Numerous extensions of this theorem to Finsler surfaces have been given, but a generalization comprising the two central features of the theorem has not been found. We will show that it *does not exist* after exhibiting these features.

The simplest form of the theorem, from which all others can be derived, states in the Riemannian case that the excess $\varepsilon(T)=\alpha+\beta+\gamma-\pi$ of a geodesic triangle T homeomorphic to a disc with angles α, β, γ equals the integral of the Gauss curvature over T. If T is simplically divided into geodesic triangles $T_1, \ldots, T_r$ then $\varepsilon(T)=\sum_{j=1}^{n}\varepsilon(T_j)$. For this conclusion it is not necessary that $\varepsilon(T)$ be an integral, but merely that the angular measure have two elementary properties: the measure is additive for angles with the same vertex and a straight angle has measure π.

More generally these properties imply that the excess $\varepsilon(P)$ of a geodesic polygonal region P on a *G*-surface defined as the sum of the excesses of the triangles $T_1, \ldots, T_r$ in a simplicial division is independent of the division:

$$\varepsilon(P)=\sum_{i=1}^{v}\varepsilon(T_i)=2\pi\,\chi(P)-\sum_{i=1}^{k}(\pi-\beta_j)$$

where $\chi(P)$ is the Euler characteristic of P and $\beta_1, \ldots, \beta_k$ are the angles of the boundary of P (see *G*, p. 283).

Thus we obtain an additive function on geodesic polygonal regions, which will behave like an integral only if it can be extended to a completely additive set function on the Borel sets. The transition to regions with curved boundaries and replacing $\Sigma(\pi-\beta_i)$ by an expression corresponding to the integral of the geodesic curvature over the boundary, is a rather simple, but depending on the generality desired, more or less laborious technical matter. This has been carried out for a definite angle and quite general regions by Alexandrov and Zalgaller ⟨1, see in particular, p. 253⟩. The method applies to any angle satisfying our two conditions which is continuous, i.e., depends continuously on the vertex and the legs (see *G*, p. 278).

For brevity *we denote as Bonnet angle a continuous angular measure for which the function* $\varepsilon(P)$ *can be extended to a completely additive function on the Borel sets.*

The existence of a Bonnet angle is the first essential feature of a true Gauss-Bonnet Theorem.

The second is the *universality of the angular measure* or that it is determined by the metric in the tangent space. To make this precise we restrict ourselves to the class $\mathfrak{C}$ of all G-surfaces which are everywhere continuously differentiable and regular in the sense of Section 5 and require: if for two points p, p' on two (possibly coinciding) G-surfaces in $\mathfrak{C}$ an isometry ϕ of the normal (Minkowski) tangent plane at p on that at p' with $\phi p = p'$ exists then the measure of any angle A with vertex p equals that of its image ϕA [8].

Universal angles exist for $\mathfrak{C}$, for example, the angles proportional to the area of the segment or the length of the arc of the unit circle of the normal Minkowskian geometry at p intercepted by an angle with vertex p. A universal angle is obviously invariant ander an isometry of an open set of one G-surface on an open set of another.

A true Gauss-Bonnet Theorem would be equivalent to the existence of a Bonnet angle universal for $\mathfrak{C}$. Such an angle would be universal for every sub-class of $\mathfrak{C}$. We will see that a universal Bonnet angle does not exist even for the sub-class $\mathfrak{C}_\infty$ consisting of the Finsler surfaces of class C^∞ in $\mathfrak{C}$. The proof proceeds as follows.

We show that a quasi-hyperbolic plane (see below) possesses, under a minimum of smoothness hypotheses, a unique Bonnet angle which is invariant under motion, and that this angle is not universal for the class $\mathfrak{H}_\infty$ of all quasi-hyperbolic planes in $\mathfrak{C}_\infty$.

A *quasi-hyperbolic plane* may be defined in different ways, for example, as a straight plane which possesses all translations along two lines L, L_1, where with suitable orientations L^+ is an asymptote to L_1^+, but L is not parallel to L_1. Then the asymptotes to L^+ are asymptotes to each other and all translations along each of these exist. They are called the *distinguished geodesics* and their common limit circles the *distinguished limit circles.* The quasihyperbolic planes can be characterized as the G-surfaces which have a nonabelian, simply transitive group of motions or as the G-planes with a group of motions isomorphic to the conformal maps of the unit disc on itself which leave one point on the boundary fixed [9].

[8] This requirement alone determines angular measure in the Riemannian case, see G p. 276. The resulting angle is, of course, a Bonnet angle.

[9] See G Sections 51, 52. For the material which follows compare Busemann [7]; the complete reference, not available when G appeared, is Rendiconti Circ. Mat Palermo Ser II, vol 4 (1955), 1–14. Quasi-hyperbolic geometry is a most valuable source for various unexpected phenomena (l c.). Here we discuss only the above mentioned fact and one which corroborates a statement made in Section 5.

Replacing the unit disc by the upper half plane $y>0$ of an (x, y)-plane and taking ∞ as the fixed point the group takes the form

$$\Gamma:\ x'=\alpha x+\beta, \qquad y'=\alpha y, \qquad \alpha>0, \beta \text{ real}.$$

A quasi-hyperbolic metric in $y>0$ with Γ as group of motions has a line *element of the form*

$$ds=y^{-1}F(dx, dy),$$

where $F(x, y)$ *is defined for all* x, y *and satisfies the conditions:* $F(x, y)>0$ *for* $(x, y)\neq(0,0)$, $F(kx, ky)=|k|F(x, y)$ *for real* k, $F(x, y)$ *is convex and differentiable for* $(x, y)\neq(0,0)$, *so that the curve* K: $F(x, y)=1$ *is convex and differentiable, finally, a tangent of* K *parallel to the* x*-axis touches* C *in one point only.*

Any ds with these properties yields a quasi-hyperbolic plane, in particular a G-space. Obviously there are K of class C^{∞} which are not strictly convex and satisfy these conditions. For such K the corresponding integrand $y^{-1}F(dx, dy)$ is nowhere regular. This confirms an assertion in Section 5 and shows that regularity is not a natural requirement for G-spaces.

Different $ds=y^{-1}F_i(dx, dy)$ $(i=1,2)$ lead to isometric G-planes if, and only if, the curves $F_i(x, y)=1$ are affinely equivalent. Therefore we may *normalize* F such that C passes through $(\pm 1, 0)$ and the tangents of C at these points are parallel to the y- resp. x-axis. The normalized F are in one-to-one correspondence with the nonisometric quasi-hyperbolic planes.

To find the *geodesics for a given* F we form the polar reciprocal of K and revolve it through $\pi/2$ obtaining a strictly convex curve K^*: $F^*(x, y)=1$ (where $F^*(kx, ky)=|k|F^*(x, y)$). The geometric relation between the Minkowski metrics in the (x, y)-plane given by $F(x-x', y-y')$ and $F^*(x-x', y-y')$ is this: *an (ordinary) line* L *is perpendicular to* L' *with respect to* F *if, and only if,* L' *is perpendicular to* L *with respect to* F^*.

The curve K^* is differentiable at its intersections with the x-axis because a tangent parallel to the x-axis touches K in one point only. If F is normalized then so is F^* in the sense that K^* passes through $(\pm 1, 0)$ with tangents parallel to the y-axis and through $(0, \pm 1)$ with supporting lines parallel to the x-axis.

The geodesics for $ds=y^{-1}F(dx, dy)$ are the intersections with $y>0$ of the curves $F^*(x-a, y)=k$ (>0) and their tangents at the intersections with the x-axis. The latter are the distinguished geodesics. The distinguished limit circles are the lines $y=k>0$. To a segment on K there corresponds a corner of K^*, as might have been expected from well known general results of Carathéodory ⟨2, Part a⟩.

Assume that a Bonnet angle invariant under Γ exists. The angle defines a completely additive set function $\varepsilon(M)$ on all Borel sets which extends the excess $\varepsilon(P)$ of polygonal regions. On the other hand the hyperbolic area in $y>0$ given by $\alpha(M)=\iint_M y^{-2}\,dx\,dy$ is also invariant under Γ, which is simply transitive, so that $y>0$ can be identified with the group space. *The theory of Haar measure shows that* $\varepsilon(M)=k'\alpha(M)$ with constant k'.

Normalize F and consider the distinguished geodesic $L: x=0$ and the geodesic L_y ending at $(a,0)$ $(a>0)$ and passing through $(0,y)$. Denote by α_y the measure of the convex angle formed by the rays on L and L_y beginning at $(0,y)$ and ending on the x-axis. If $y_1>y$ then $\alpha\, y_1<\alpha_y$. For, the motion $x'=\alpha x$, $y'=\alpha y$ $(\alpha=y y_1^{-1})$ takes $(0,y_1)$ into $(0,y)$ and $(a,0)$ into $(\alpha a,0)$. Therefore $\alpha=\lim_{1\to\infty}\alpha_y$ exists, moreover $\alpha=0$ because for any $0<a_1<a$ the motion $x'=a a_1^{-1}x$, $y'=a a_1^{-1}y$ takes $(a_1,0)$ into $(a,0)$ and the measure of the angle with vertex $(0,y)$ to $(0,0)$ and $(a_1,0)$ tends to 0 when $a_1\to 0$ (the angular measure is continuous). Similarly

$$\alpha_y\to\pi \quad \text{for } y\to 0+.$$

For any fixed nondistinguished geodesic, say K^* in $y>0$, consider the convex angle $A_c(-1<c<1)$ formed by the ray $x\geqq c$, $y>0$ of K^* and the subray $x\geqq d$ of $x=c$ beginning at the intersection (c,d) of $x=c$ with K^*. Using translations $x'=x$, $y'=y+\beta$ in Γ we deduce from the preceding observations that the measure β_c of A_c satisfies $\beta_c\to 0$ for $c\to -1+$ and $\beta_c\to\pi$ for $c\to 1-$.

In polar coordinates (r,ϕ) let K^* be given by $r=g(\phi)$ $(0\leqq\phi\leqq 2\pi;$ $g(\phi+\pi)=g(\phi))$ and in terms of x and y for $y\geqq 0$ by $y=f(x)$. Our discussion implies that the angle sum in the geodesic triangle with vertices $(-1+\varepsilon, f(-1+\varepsilon))$, $(-1+\varepsilon,\varepsilon^{-1})$, $(-1+\varepsilon,\varepsilon^{-1})$, $(1-\varepsilon, f(1-\varepsilon))$ tends to 0 when $\varepsilon\to 0+$, and hence its excess to $-\pi$. This shows that k' in $\varepsilon(M)=k'\alpha(M)$ is negative.

The domain M_c: $\{-1\leqq x\leqq c;\ y\geqq f(x)\}$ has

$$\alpha(M_c)=\int_{-1}^{c}\int_{f(x)}^{\infty} y^{-2}\,dx\,dy=\int_{-1}^{c} f^{-1}(x)\,dx.$$

This will be finite if K^* has at $(-1,0)$ a finite nonvanishing curvature. Since we want to produce Finsler spaces of class C^∞ we assume the Legendre Condition which amounts to requiring that K has everywhere nonvanishing curvature and is of class C^∞. The same holds then for K^*. We know that $\varepsilon(M_c)=-\pi$. Since $\beta_1=\pi$ we find

$$-k'=k=\pi\Big/\int_{-1}^{1} f^{-1}(x)\,dx,\qquad \beta_c=k\int_{-1}^{c} f^{-1}(x)\,dx.$$

If $(c, f(c)) = (g(\alpha), \alpha)$ then

$$\beta_c = k \int_{\pi}^{\alpha} (g'(\phi)\cos\phi - g(\phi)\sin\phi)(g(\phi)\sin\phi)^{-1}\, d\phi$$

$$= k(\pi - \alpha) - k \int_{\alpha}^{\pi} g'(\phi)\, g^{-1}(\phi) \cot\phi\, d\phi.$$

For $g(\phi) = 1$ we find the hyperbolic value $\pi - \alpha$.

The measure of a general angle is obtained from β_c by additivity. We want to show that *this angle is not universal even for the quasi-hyperbolic planes.*

For this purpose we consider with the same K the quasi-hyperbolic plane in $x > 0$ defined by

$$ds_1 = x^{-1} F(dx, dy).$$

Since multiplying all distances in a Minkowski geometry by the same factor yields an isometric space, the local Minkowskian geometries at all points in both quasi-hyperbolic planes is isometric to $F(x - x', y - y')$.

Choose $g(\phi)$ such that $g(\phi) = 1$ for $0 \leqq \phi \leqq \pi/2$. We evaluate the measure of A_0 in the two metrics.

Putting $Q(u) = \pi(\pi/2 - u)(\pi - u)^{-1}$ we find

$$\beta_0 = Q\left(\int_{\pi/2}^{\pi} g'(\phi)\, g^{-1}(\phi) \cot\phi\, d\phi\right)$$

and for ds_1, the measure is

$$Q\left(\int_{-\pi/2}^{0} g'(\phi)\, g^{-1}(\phi) \tan\phi\, d\phi\right) = Q\left(\int_{\pi/2}^{\pi} g'(\phi)\, g^{-1}(\phi) \tan\phi\, d\phi\right).$$

These two measures differ for suitable choices of $g(\phi)$ in $\pi/2 < \phi < \pi$. Thus we proved:

(1) *Theorem. There is no universal Bonnet angle for all G-surfaces in* $\mathfrak{C}$ *or all G-surfaces in* $\mathfrak{C}_\infty$.

It is conceivable, but quite improbable, that each G-surface in $\mathfrak{C}_\infty$ possesses a Bonnet angle, however the problem has not been investigated.

(1) justifies the partial generalizations of the Gauss-Bonnet Theorem given by various authors. Among these the extension of the higher-dimensional theorem by Lichnerowicz ⟨1⟩ seems especially significant.

17. Various Aspects of Conjugacy

We conclude the chapter on geodesics by reporting on some investigations related to the distribution of conjugate points.

A G-space R has nonpositive curvature if, locally, points a, b, c and the midpoints b' of $T(a, b)$ and c' of $T(a, c)$ satisfy

$$2b'c' \leqq bc.$$

Under the assumption of domain invariance this implies the absence of conjugate and of focal points in R, see G, pp. 163, 240, 254, as well as the validity of all the results on Riemann spaces with nonpositive curvature.

a) Spaces with Curvature $\geqq K > 0$

Requiring $2b'c' \geqq bc$ leads to spaces with nonnegative curvature, however, the most interesting results in the Riemannian case concern spaces where the sectional curvature is bounded below by a positive constant K.

How to define this concept for G-spaces was shown by Kann ⟨1⟩: Let a, b, c be in $S(p, \rho(p)/8)$. For $b' \neq c'$ (defined as above) consider the segment T of length $\rho(p)$ which has the same midpoint, and contains $T(b', c')$. A foot f of a on any segment which contains T lies in the interior of T. Put

$$p(a; b, c) = aT = af.$$

Then *Kann defines curvature* $\geqq K$ ($\leqq K$) by $2b'c' \geqq bc + Kp^2(a; b, c)$ resp. $2b'c' \leqq bc + Kp^2(a; b, c)$ and proves that this agrees for Riemann spaces with the usual concepts.

According to Bonnet (1855) a compact Riemannian surface with curvature $\geqq K > 0$ has diameter at most $\pi K^{-\frac{1}{2}}$. Hopf and Rinov ⟨1⟩ extended this to complete surfaces. This was generalized to n dimensions by several mathematicians, first, apparently, by Schoenberg ⟨1⟩. Kann proves:

A two-dimensional G-space in which, locally, the circles are convex and the curvature is $\geqq K > 0$ *has diameter at most* $2\pi K^{-\frac{1}{2}}$.

The assumption that small circles be convex is acceptable because it is satisfied by Finsler spaces, see Whitehead [1], also G, p. 162. The restriction to two dimensions is objectionable. It becomes necessary if one tries to follow closely the method used in G for $K \leqq 0$, although there dimension 2 does not enter. Under adequate differentiability hypotheses (guaranteeing that the local behavior of the geodesics resembles sufficiently that of straight lines) the hypothesis can be eliminated.

Owing to the factor 2 Kann's theorem does not contain Bonnet's. However, Kann shows that *the 2 may be omitted if perpendicularity of segments is symmetric*, which is the case in Riemann spaces.

The factor can also be avoided by using transversality instead of perpendicularity. Consider a segment T' through a to which the extension T of $T(b', c')$ is perpendicular, i.e., T' is transversal to T. In case T'

is unique (which amounts to a differentiability hypothesis) put $T' \cap T = g$ and $ag = t(a; b, c)$. Then:

In a two-dimensional G-space let, locally, the circles be convex and the transversal unique. If

$$2b'c' \geqq 2bc + Kb'c't^2(a; b, c), \qquad K > 0$$

then the diameter of R is at most $\pi K^{-\frac{1}{2}}$.

Completely different approaches by Auslander ⟨1⟩ and Moalla ⟨1⟩ are mentioned in the last chapter.

All these investigations rest on Jacobi's equation for conjugate points (because of weak assumptions Kann uses a difference equation) and show that the first conjugate point to a given point on any geodesic has at most distance $\pi K^{-\frac{1}{2}}$ (resp. $2\pi K^{-\frac{1}{2}}$) from the point.

b) The Divergence Property

We now turn to a case where the absence of conjugate points can be proved. Generalizing a definition of *G*, p. 230 we may say that a (non-compact) *G*-space has the divergence property if for any two distinct rays R_1, R_2 with the same origin the distances xR_2 and xR_1 tend to ∞ when x traverses respectively R_1 or R_2.

The question whether the universal covering space of a compact space has the divergence property if it is straight, is important for the existence of transitive geodesics (see *G*, Section 34) and was therefore raised in Problem (34) *G*, p. 406. It is still open, even for surfaces. However, Zaustinsky ⟨2⟩ noticed that the converse holds in the latter case:

(1) *If the universal covering space of a compact G-surface is a plane with the divergence property, then it is straight.*

There are two cases, the euclidean forms or the torus and one-sided torus, which are treated by the method mentioned at the end of Section 12, and the hyperbolic forms. For these we deduce (1) from the following more general theorem:

(2) *Let xy be a metrization of the hyperbolic plane H^2* (with curvature -1) *as a G-space P and let numbers $0 < \alpha \leqq \alpha'$ exist such that $xy < \alpha$ or $h(x, y) < \alpha$ imply, respectively, $h(x, y) < \alpha'$ or $xy < \alpha'$. If P has the divergence property then it is straight.*

As in Section 12 we realize H^2 as the Klein Model in the interior of the unit circle K in E^2 and conclude from (12.1) that for $x \in P$ and $p_\infty \in K$ an $R(x, p_\infty)$ and for two distinct points q_∞, r_∞ on K a straight line $L(p_\infty, q_\infty)$ in P exists.

The set $F = \bigcup_{p_\infty \in K} R(x, p_\infty)$ is a closed subset of $P \cup K$ containing K. If F is a proper subset of $P \cup K$ then for some p_∞ two distinct rays

$R(x, p_\infty)$ exist. As in the proof of G (34.8) one finds that the convex domain bounded by two such rays cannot contain a diverging pentagon (in P). On the other hand, it follows from (12.1) that the hyperbolic η-neighborhood of a hyperbolic diverging pentagon contains a diverging pentagon in P, so that for a suitable η every $S(q, \eta)$ in P contains such a pentagon. The divergence property therefore yields (compare G (34.9)) that $R(x, p_\infty)$ is unique and hence $F = P \cup K$. This means that a given $y \neq x$ lies on a ray with origin x, so that P is straight.

The argument using the pentagon requires $n=2$ and also that P be a G-space rather than merely a space with an intrinsic metric as in (12.1). The symmetry of the distance is not essential.

c) Conjugate Points to Rays

Conjugacy to points at infinity has not been studied in the calculus of variations or differential geometry. The analogue in a G-space to a point at infinity is a maximal set of rays such that a co-ray to one is a co-ray to all others[10], but it proves more convenient to operate with individual rays.

The merit of uncovering this interesting topic belongs to Nasu ⟨1, 2, 3⟩. Unfortunately some of his assertions are false and many of his proofs are incomplete or contain mistakes. G. M. Lewis[11] took Nasu's ideas up, provided valid proofs, frequently generalized the results materially and carried the investigation farther. A brief report on the theory as hitherto developed will, hopefully, convince the reader, that here is a field with many opportunities for worthwhile research. It is always assumed that the space is a noncompact G-space, so that each point is the origin of at least one ray.

The ray B with origin p is a *co-ray* from p to the ray A, if sequences of points $p_\nu \to p$ and $x_\nu \in A$ with $p x_\nu \to \infty$ exist such that suitable segments $T(p_\nu, x_\nu)$ tend to B. We write this briefly as $B \underset{p}{\succ} A$ or $B \succ A$ when knowing p is not relevant. Obviously:

(3) *For given p and A a co-ray from p to A exists. $B \succ A$ implies $B \succ A'$ for $A' \subset A$ or $A' \supset A$.*

(4) *If $B_\nu \succ A$ $(\nu = 1, 2, \ldots)$ and B_ν converges to a ray B then $B \succ A$.*

Less trivially, see G (22.19):

(5) *If $B \underset{p}{\succ} A$ and $g \in B - \{p\}$ then the co-ray from g to A is unique and is a subray of B.*

---

[10] See G. Sections 22, 23. The definition and simplest properties are also found below.

[11] Cut loci of points at infinity, Dissertation University of Southern California 1970. An abbreviated version will be published.

Here arises an important question:

(6) *Do $B' \supset B$ and $B \succ A$ imply $B' \succ A$?*

The problem is related to those mentioned at the end of Section 4 and was raised in Busemann [3], but is still open. Two examples in Nasu ⟨2, 3⟩ claiming to provide a negative answer to (6) are false.

With the obvious definitions of *symmetry* (if $B \succ A$ then $A \succ B$) and *transitivity* ($B \succ A$ and $C > B$ imply $C \succ A$ co-ray relation we have:

(7) *Transitivity implies symmetry.*

For, let $B \succ A$. Let r be any point of A different from its origin g and $C \underset{r}{\succ} B$. Then $C \succ A$, hence $C \subset A$ and $A \succ B$ follows from (4) for $r \to g$.

(8) *If the co-ray relation is symmetric (or transitive) then $B' \supset B$ and $B \succ A$ imply $B' \succ A$.*

For $A \succ B$ and, by (3), $A \succ B'$ hence $B' \succ A$. Since (6) and the symmetry of the co-ray relation are not naturally related the following seems a more adequate form of symmetry: If $B \succ A$ then $A' \succ B$ for a suitable $A' \subset A$. But a similarly satisfactory formulation of transitivity has not been found. Even *the more restricted form of symmetry, and hence transitivity, do in general not hold*, see *G*, p. 139 and Section 15 here.

For a given A and $B \succ A$ consider the union of all rays which contain B and are co-rays to A. Because of (4) this is either a straight line or a ray. In the first case, we call the line with the orientation for which B is a positive subray, an *asymptote* to A; in the second a *maximal* co-ray to A and its origin a co-point to A [12]. If the maximal co-ray is also the maximal ray containing B we call it *terminal.*

A co-point to A is the natural concept of a conjugate (or minimum) point to the point at infinity on A.

Denote by $C(A)$ the set of all co-points to A. The points which are origins of two or more co-rays to A form, because of (5), a subset $C_2(A)$ of $C(A)$. Put $C_1(A) = C(A) - C_2(A)$. The complement of $C_2(A)$ consists of those points from which the co-ray to A is unique.

(9) *The complement $C'(A)$ of $C(A)$ has no bounded component and no compact subset of $C(A)$ separates the space.*

For if $p \in C'(A)$ and $B \underset{p}{\succ} A$ then $B \cap C(A) = \emptyset$ by the definition of co-point, so that the component of $C'(A)$ defined by p contains B.

If a compact subset K of $C(A)$ existed which separates the space, then A with the possible exception of its origin, would lie in one compo-

[12] The terminology in Busemann [3] and Nasu ⟨1⟩ is different. There also maximal co-rays are called asymptotes and Nasu uses "asymptotic conjugate point" instead of co-point.

nent of the complement of K. Let p lie in an other. A segment $T(p, x_\nu)$ with $x_\nu \in A$ and $p x_\nu \to \infty$ intersects K in some point y_ν. For a subsequence $\{\mu\}$ of $\{\nu\}$ the segment $T(p, x_\mu)$ converges to a co-ray B from p to A and y_μ to a point $y \in K$. But $y \in B \cap C(A)$ is impossible.

The set $C(A)$ *need not be closed*, see Nasu ⟨1, 3⟩, where also the following two facts are proved.

(10) *If $C(A)$ is nonempty and compact then no asymptote to A exists.*

(11) *If $C(A)$ contains an isolated point g then $C(A) = \{g\}$ and the space is the union of rays with origin g* (and hence simply connected). *These rays are co-rays to each other. The limit spheres with these rays as central rays are the ordinary spheres $K(g, \rho)$.*

An argument similar to that used in the proof of (4.11) yields

(12) *If all spheres $K(x, \rho)$ $(0 < \rho < \rho(x))$ are contractible* (hence if $\dim R < \infty$) *then $C_2(A)$ is dense in $C(A)$. Therefore, if $C_2(A) = \emptyset$ then $C(A) = \emptyset$ and each point lies on an asymptote to A.*

The second statement is the analogue to (4.12f) for the case where p is at infinity.

The deeper results hitherto obtained concern surfaces. A tube on a G-surface is a closed domain bounded a simple closed geodesic polygon and homeomorphic to a disc punctured at one (interior) point, see G, Section 43.

(13) *If the universal covering space $\bar{R}$ of the G-surface R is straight and a subray of the ray A lies in a tube, then the number $m(p, A)$ of co-rays from p to A is finite for any p and a maximal co-ray is terminal.*

(14) *If, in addition to the hypothesis of* (13), *the co-ray relation in R is transitive, then $C_1(A) = \emptyset$ (or $m(p, A) \geqq 2$ for $p \in C(A)$) and $C(A)$ is closed.*

Each point $p \in C(A)$ has a neighborhood V whose closure $\bar{V}$ can be mapped topologically on a disc B such that p goes into the center of B and $C(A) \cap \bar{V}$ into the union of $m(p, A)$ radii of B.

The hypothesis that a subray of A lies in a tube is essential, see the example in G, pp. 265–266. If R has finite connectivity then every ray has a subray which lies in a tube.

Denote as *$C(A)$-path with origin p* a topological map ϕ of $t \geqq 0$ into $C(A)$ with $\phi(0) = p$ and $\phi(0)\,\phi(t) \to \infty$ for $t \to \infty$.

If every closed Jordan curve decomposes R then $C(A)$ cannot contain such a curve by (9), and we conclude from (14) that each arc on $C(A)$ with p as endpoint can be extended to an, in general not unique, $C(A)$-path with origin p and that each $p \in C(A)$ is the origin of at least two $C(A)$-paths which have only p in common.

The *co-ray distance* in a G-space is said to be *bounded* if for $B \underset{p}{\succ} A$ a sequence $y_\nu \in B$ with $p y_\nu \to \infty$ exists for which $\{y_\nu A\}$ is bounded. (Remember that in a straight space with convex capsules both yA and xB are bounded when $B \succ A$, G (37.1).)

(15) *If R is a G-space of finite connectivity with a straight universal covering space in which the co-ray distance is bounded and the co-ray relation is transitive, then this relation is transitive in R.*

The proofs of (13), (14), (15) by Lewis are quite involved. Nasu ⟨2, 3⟩ states these facts for R with nonpositive curvature, finite connectivity and differentiable small circles. However, his arguments are often dubious. Most of the following more precise information is due to Lewis:

(16) *Under the assumptions of (15) let R be topologically a compact surface of genus g punctured at k points, and put $\bar{g}=2g$ or $\bar{g}=g$ according to whether R is orientable or not. Let $h(A)$ and $j(A)$ be respectively the number of components of $C(A)$, and the number of points on $C(A)$ with $m(p, A)>2$. Then*

$$m(p, A) \leqq k+\bar{g}, \quad h(A) \leqq k+\bar{g}-1, \quad j(A) \leqq k+\bar{g}-2.$$

V. Motions

Much work has been done on spaces with large groups of motions, but very little on conditions under which the number of motions is finite. The best known classical result in this direction states that a compact Riemann space with negative curvature has a finite group of motions. Section 18 discusses questions of this type and in particular extends the mentioned theorem in several ways.

The last part of G gives a nearly complete treatment of spaces with pairwise transitive groups of motions. It would seem natural to complete this work by discussing the remaining case of noncompact even dimensional spaces, for which proofs appeared soon after 1955. The intervening development has confronted us with a dilemma to which we have not found a satisfactory answer. It appears that the theory of spaces with this mobility is best considered in the context of transformation and Lie groups, with the metric entering only at the last stage. We explain this new theory without proof and settle in the compact case a closely related problem which does typically concern G-spaces, namely: when is the group of motions transitive on the geodesics?

18. Finite and One-Parameter Groups of Motions

Under various hypotheses it can be shown that the group of all motions of a G-space is finite. We begin with a very simple case[1]:

(1) *A compact manifold with nonvanishing Euler characteristic which is metrized as a G-space without conjugate points possesses only a finite number of motions.*

We mean here a topological manifold which is a simplicial complex (or only a combinatorial manifold). Because no general agreement on the sign of the characteristic χ exists we mention for (2) below that we use $\chi = \Sigma(-1)^k \alpha^k$, if α^k is the number of k-simplices.

If (1) were false then the group of all motions of R, which is by G (52.3) a Lie group, would contain a one-parameter subgroup Φ_t $(-\infty < t < \infty, \Phi_{t_1} \Phi_{t_2} = \Phi_{t_1+t_2})$. At least one orbit of Φ_t is a geodesic $x(t)$, i.e., $\Phi_t x(0) = x(\alpha t)$ $(\alpha > 0)$, see G (52.2). Because $\chi \neq 0$ there is a point f

[1] Most results of this section are taken from Busemann ⟨6, 7⟩.

fixed under all Φ_t, see Seifert-Threlfall [1, p. 251], where the opposite sign is chosen for χ.

We pass to the universal covering space $\bar{R}$ of R. For a given point $\bar{f}$ in $\bar{R}$ there is a group of motions $\bar{\Phi}_t$ of $\bar{R}$ over Φ_t with $\bar{f}$ as fixed point and this has a geodesic $\bar{x}(t)$ over $x(t)$ as orbit. By hypothesis $\bar{x}(t)$ is a straight line, so that $\bar{f}\,\bar{x}(t)\to\infty$ for $t\to\infty$, but $\bar{f}\,\bar{x}(t)=\bar{f}\,\bar{x}(0)$ since $\bar{\Phi}_t$ is a motion.

Using (1) we prove

(2) *A compact G-surface with negative Euler characteristic has a finite total group of motions.*

We start the proof as in (1) obtaining f, $\bar{R}$, $\bar{\Phi}_t$, $\bar{f}$. Then $\bar{R}$ is, topologically, a plane because $\chi<0$. The $\bar{\Phi}_t$ are rotations about $\bar{f}$. In the same way we obtain the rotations about $\bar{f}_1 \neq \bar{f}$ over f. Now $\bar{R}, \bar{f}, \bar{f}_1$ satisfy the hypotheses for R, p, q in (14.4) and we conclude that the metric of $\bar{R}$ is euclidean or hyperbolic, hence $\bar{R}$ is straight and the assertion follows from (1).

Results covering odd-dimensional spaces will be found below. We need the following stronger form of statements in G, Section 32:

(3) *In a straight space R let a compact subset C and a group Γ of motions exist such that any point of R can be mapped into C by a motion in Γ. If the motion $\phi \neq 1$ (not necessarily in Γ) commutes with all elements of Γ, then the distance $x\phi x$ is independent of x. Every point lies on an axis of ϕ and the axes of ϕ are parallel.*

For, if y is given, choose $\psi\in\Gamma$ with $\psi y\in C$. Then $y\phi y=\psi y\psi\phi y=\psi y\phi\psi y\leq \max_{x\in C} x\phi x$. Therefore $y\phi y$ attains its maximum (at a point of C). The rest follows from G (32.4, 5).

Our first application of (3) is an improvement of G (32.11):

(4) *Theorem. Let R be compact and without conjugate points. The closed geodesics belonging to an element ϕ ($\neq 1$) in the center of the fundamental group Δ of R are simple, have the same length and cover R simply.*

For, ϕ forms in Δ a class of conjugate elements and defines therefore a class $[\phi]$ of freely homotopic curves in R, see G, p. 208. Because R is compact, $\bar{R}$ contains a fundamental set with a compact closure C. Any point of $\bar{R}$ can be moved into C by an element of Δ (realized as the group of motions of $\bar{R}$ lying over the identity of R). Thus the hypotheses of (3) are satisfied by ϕ and Δ. The images of the axes of ϕ in R are closed geodesics in $[\phi]$. They cover R since the axes of ϕ cover $\bar{R}$. If two such geodesics had a common point q or one had a multiple point q, then two different axes of ϕ would pass through a point $\bar{q}$ over q. The geodesics have the same length because $x\phi x$ is independent of x.

(5) *Corollary. The fundamental group of a compact G-space with domain invariance and strictly convex capsules*[2] *does not have a center.*

This strengthens G (39.5) and follows from (4), G (38.2) and (40.2) or (37.1).

A second implication of (3) is:

(6) *Let Φ_t be a one-parameter group of motions of a compact G-space* R $(\dim R > 1)$ *without conjugate points. If $\bar{\Phi}_t$ is a group of motions over Φ_t of the universal covering space $\bar{R}$ of R, then the orbits of $\bar{\Phi}_t$ are parallel straight lines and $\bar{x}\bar{\Phi}_t\bar{x}$ is independent of $\bar{x}$ for each t. Hence, $x\Phi_t x$ is independent of x for small $|t|$ and the orbits of Φ_t are simple geodesics. Φ_t has no fixed points for small $|t|>0$.*

The $\bar{\Phi}_t$ commute with the elements of the fundamental group Δ of R, compare G (28.11). If C is defined as in the proof of (4), then (3) applies and (6) follows readily.

(7) *Corollary. A compact G-space with strictly convex capsules and domain invariance possesses only a finite number of motions.*

This follows again from G (37.1) and (38.2) and contains the well known fact that a compact Riemann space with negative curvature admits only a finite number of motions. We can prove more and obtain a result closely related to a theorem of Bochner ⟨1⟩ which states that the number of motions of a compact Riemann space with negative Ricci curvature is finite. We show first:

(8) *Theorem. Let the compact Riemann space R without conjugate points (and dimension greater than 1) possess a one-parameter group of motions Φ_t. Then the universal covering space $\bar{R}$ of R is the product of a t-axis and a straight space Q and the line element of $\bar{R}$ (hence of R) has the form*

$$ds^2 = dq^2 + a\,dt^2,$$

where dq^2 is the line element of Q and a is constant.

We construct $\bar{\Phi}_t = \psi_t$ as above. The orbits of ψ_t are straight lines by (6). The limit spheres with the oriented orbits as central rays (see G, Section 23) are of class C^1 (at least, under the usual differentiability assumption) and intersect the orbits orthogonally. Therefore they do not depend

[2] Remember that a Riemann space with negative curvature has strictly convex capsules, see G(41.7).

on the central ray, i.e., if L_x^+, L_y^+ are the orbits $\psi_t x$, $\psi_t y$ (omitting bars for points in $\bar{R}$) traversed with increasing t, then $K_\infty(L_x^+, p) = K_\infty(L_y^+, p)$ for any p.

We show that $K_\infty(L_x^+, x)$ consists of lines orthogonal to L_x^+ at x. Let H be any line orthogonal to L_x^+ at x. Then ψ_t is orthogonal to L_x^+ at $\psi_t x$ for any t. If $z \in H$, then $x\,\psi_t x = z\,\psi_t z$ by (6) and also $x\,z = \psi_t x\,\psi_t z$. When t varies from $-\infty$ to ∞ the lines $\psi_t H$ traverse a manifold M_H homeomorphic to a plane and the $\psi_t H$ intersect the orbit $L_z^+ = \psi_t z$ at constant angle ($z \in H$). Therefore the intrinsic metric on M_H is euclidean by the Gauss-Bonnet Theorem. The distance relations now show that the $\psi_t H$ is orthogonal to L_z^+, hence $H \subset K_\infty(L_x^+, x)$.

If $y \in K_\infty(L_x^+, x)$ then $K_\infty(L_y^+, y) = K_\infty(L_x^+, x)$ shows that $K_\infty(L_x^+, x)$ also consists of lines through y. Therefore $K_\infty(L_x^+, y)$ is a flat space Q. The euclidean character of the metric in the M_H implies that the line element has the stated form.

(9) *Corollary. A compact Riemann space without conjugate points and with dimension greater than* 1 *admits only a finite number of motions, unless every point p lies on a lineal element* L_p *such that the sectional curvature of every plane element through* L_p *vanishes.*

Finally we want to strengthen (7) in a different direction and prove for this purpose a fact which is of independent interest. It uses the original interpretation Δ_p of the fundamental group, where the elements are classes of homotopic curves beginning and ending at some fixed point p.

(10) *Theorem. If the G-space R has a finitely generated fundamental group* Δ *and* Φ_t *is a one-parameter group of motions of R which has a closed orbit, then a suitable (positive) multiple of this orbit commutes with all elements of* Δ.

"Closed orbit" means that for a suitable point p and some fixed $k>0$ the orbit $\Phi_t p$ satisfies $\Phi_{t+k} p = \Phi_t p$ for all t. We denote the closed curve $\Phi_t p$ $(0 \leq t \leq k)$ by α and realize Δ as Δ_p. The assertion is trivial for $\alpha \sim 0$ (which includes the possibility $\Phi_t p \equiv p$) or if α generates Δ_p. Assume that neither of these cases enters and take any $\beta \in \Delta_p$ not in $\{\alpha^\nu\}$. We can realize β as a shortest curve in its homotopy class, but need only that β may be chosen as a rectifiable curve (so that (10) is valid for much more general spaces than G-spaces).

$\Phi_{mk}\beta$ has the same length as β is obtained from β by the deformation $\Phi_t \beta$ $(0 \leq t \leq mk)$ under which p traverses α^m. Therefore

$$\Phi_{mk}\beta \sim \alpha^m \beta \alpha^{-m}.$$

There is only a finite number of homotopy classes in Δ_p which can be realized by curves whose lengths stay below a given bound. Therefore with a suitable $n \geqq 1$

$$\beta \sim \Phi_{nk} \beta \sim \alpha^n \beta \alpha^{-n}.$$

By hypothesis Δ_p can be generated by a finite number of elements $\alpha, \beta_1, \ldots, \beta_j$. Choose $n_i \geqq 1$ such that

$$\beta_i \sim \alpha^{n_i} \beta_i \alpha^{-n_i}, \qquad i=1, \ldots, j,$$

and let n_0 be the least common multiple of $n_1, \ldots, n_j$. Then $\beta_i \sim \alpha^{n_0} \beta_i \alpha^{-n_0}$ for all i. Since α and the β_i generate Δ_p we have

$$\gamma \alpha^{n_0} = \alpha^{n_0} \gamma \qquad \text{for all } \gamma \in \Delta_p.$$

This and the preceding facts yield:

(11) *Theorem. If in a compact G-space R without conjugate points (of dimension greater than 1) a nontrivial free homotopy class K of closed curves exists such that for all* $m \geqq 1$ *the shortest curve in* K^m *is unique then R admits only a finite number of motions.*

The proof is again indirect: if a one-parameter group Φ_t of motions did exist, and α is a shortest curve, hence a closed geodesic, realizing K, then $\Phi_t \alpha \sim \alpha$ would give $\Phi_t \alpha = \alpha$. Hence Φ_t either leaves α pointwise fixed or has α as orbit. The first case is impossible by (6).

In the second α^m lies because of (10) for some $m > 0$ in the center of the fundamental group, which is finitely generated since R is compact (see G (29.6)), and $\alpha^m \neq 0$ because $\bar{R}$ is straight. But then the closed geodesics homotopic to α^m would cover R and have the same length by (4), which contradicts our uniqueness hypothesis.

19. Transitivity on Pairs of Points and on Geodesics

The theory of spaces with pairwise transitive groups of motions[3] in G, Sections 54, 55 is not complete, because the relevant papers for the even-dimensional noncompact case had not yet appeared.

We do not resume the discussion here, because following an impulse of Kolmogoroff [1], Tits [1] and ⟨1, 2⟩ and Freudenthal ⟨1, see also 2⟩ have given the theory a different aspect in which *the metric results become corollaries of theorems on topological transformation groups.*

To develop this theory one would have to start *ab ovo* which is hardly justifiable in view of the mentioned papers and the part of the theory already found in G. Most arguments would be foreign to the present methods, like Section 3 on r-spaces, but, unlike Section 3, no other part of our theory requires the results. Therefore we give without proof a

[3] I.e., given points a, a', b, b' with $ab = a'b'$ there exists a motion taking a into a' and b into b'. Triplewise transitivity is defined analogously.

brief description of the new theory following Freudenthal and then turn to a closely related topic which belongs inherently to G-spaces.

Let R be a locally compact connected Hausdorff space with the following properties I *to* IV.

I. *R possesses a transitive group Γ of homeomorphisms satisfying the conditions:*

II. *Given a compact set A and a closed set B disjoint from A there is an open set $U \neq \emptyset$ such that $\phi U \cap A = \emptyset$ or $\phi U \cap B = \emptyset$ for all $\phi \in \Gamma$.*

Note that II is satisfied if R is metric and Γ a group of motions. I and II furnish a uniform structure in R invariant under Γ.

III. *Γ is complete in the sense of this structure.*

It is not known whether III can be omitted or not. For a point $p \in R$ denote by Γ_p the stabilizer of p in Γ. Because Γ is transitive Γ_p and $\Gamma_{p'}$ are isomorphic for any p, p'. The orbit of q under Γ_p is denoted by $\Gamma_p(q)$.

IV. *There are points p, q for which $\Gamma_p(q)$ separates the space.*

In a metric space with a pairwise transitive group of motions the group of all motions satisfies $\Gamma_p(q) = K(p, pq)$, which separates the space if pq is positive and smaller than the diameter of the space.

Using Yamabe's solution of the generalized Fifth Problem of Hilbert one proves that Γ is a Lie group with a finite number of components, that Γ_p is compact, that a metric invariant under Γ exists and that Γ_p is transitive on $K(p, \rho)$. The theory of representations of semi-simple Lie groups leads to a classification of the possible Γ_p and their imbedding in Γ with the following result:

(1) *Theorem. Γ is isomorphic to a closed subgroup which is contained in, and contains the identity component of, the group of motions of one of the following spaces:*

(2) *The elementary and elliptic spaces, the elliptic and hyperbolic hermitian or quaternian spaces, the elliptic and hyperbolic Cayley planes*[4].

In addition to (2) *Γ may be one of six special groups* (which we do not list).

The answer to our metric problems is derived from (1) in the following form:

(3) *Theorem. Let R be a locally compact connected metric space and Γ the group of all motions of R.*

If for some positive γ smaller than the diameter of R and any four points a, a', b, b' with $ab = a'b' = \gamma$ a motion exists which takes a into a' and b into b', then R is one of the spaces listed under (2) *with the reservation that the distance may have the form $f(xy)$, if xy is the usual distance of the space, which is distinguished by intrinsicness.*

[4] These spaces are described in *G* Section 53.

The six special cases are eliminated by applying (1) to Γ_p in (3). This is not carried out by Freudenthal, but can be extracted from Tits ⟨2⟩, where the proofs for the results announced in Tits [1] and ⟨1⟩ are found.

Using (2) one can easily obtain the elementary spaces by requiring triple-wise transitivity. But in this case the procedure in G is preferable, at least for G-spaces[5], because *an elementary approach to these spaces is obviously desirable.* This is accomplished in G via the Bisector Theorem, which is of basic importance by itself and characterizes the elementary spaces (with the exception of the circle) by the property that the bisectors (i.e., loci of points equidistant from two distinct points) are flat. *This theorem does not follow from, but yields at once, the characterization by triplewise transitivity*, and is thus the stronger result.

We now turn to the problem on G-spaces to which we alluded above. Let $x(t)$, $y(t)$ represent geodesics in a G-space R and let X, Y be the point sets carrying them. If a motion ϕ of R exists with $\phi X = Y$, then there are also numbers $\eta = \pm 1$ and α such that $\phi x(t) = y(\eta t + \alpha)$ for all t, see G, p. 39. We say that ϕ takes X or $x(t)$ into Y or $y(t)$.

It is obvious that in a space with a pairwise transitive group of motions a motion exists which takes one of two given geodesics into the other[6]. The converse is true for compact spaces, the question is open in the unbounded cases except for dimensions 2 and 3:

(4) *If the group Γ of motions of a two-dimensional G-space, or of a three-dimensional G-space which is continuously differentiable and regular at some point p, is transitive on the set of geodesics, then the space is elementary or elliptic.*

(5) *Theorem. If the group Γ of motions of a compact G-space is transitive on the set of geodesics then the space is one of the compact spaces in the list* (2).

In both theorems we may assume that Γ is the group of all motions. In *the two-dimensional case* $\dim \Gamma \geqq 2$ and the assertion follows from (14.5).

In *three dimensions* and under the differentiability hypothesis the space is a topological manifold, see (5.9), the $K(p, \rho)$ $(0 < \rho < \rho(p))$ are also spheres of the normal Minkowskian tangent space at p and hence homeomorphic to S^2. The geodesics form a 4-parameter family, so that $\dim \Gamma \geqq 4$ and $\dim \Gamma_p \geqq 1$. The group Γ_p may be interpreted as a group of rotations (including, if necessary, motions which change the orientation)

[5] Actually triplewise transitivity allows us at once to reduce the metric conditions materially.

[6] The remainder of this section is taken from Busemann ⟨3⟩.

about p of the tangent space, hence as a subgroup of the rotation group of S^2, see G, p. 101.

If $\dim \Gamma_p \geqq 2$, then Γ_p is transitive on S^2 and hence contains the entire orthogonal group, see Montgomery and Zippin ⟨1⟩. If $\dim \Gamma_p = 1$ then Γ_p contains a circle group. In either case Γ_p contains the rotations about some lineal element L at p and therefore the reflection in L (see G, Section 49). This induces a reflection in each lineal element on the geodesic containing L which can be moved into any other geodesic. Therefore the reflection in a given lineal element exists, so that the space is elementary or elliptic by G (49.7).

In the *compact case* we assume that the space has dimension greater than 1. Then Γ is a Lie group (see G (52.4)) of positive dimension and hence has one parameter subgroups. We use again that such a subgroup has an orbit which is a geodesic $x(t)$, G (52.2). By adjusting the parameter in the subgroup we will have

$$x(t) = \Phi_t\, x(0) \qquad \text{for all } t.$$

For any other geodesic Y there is a motion ψ taking Y into $x(t)$ and we can choose the representation $y(t)$ of Y so that $\psi\, x(t) = y(t)$. Then

$$\psi\, \Phi_t\, \psi^{-1}\, y(0) = \psi\, \Phi_t\, x(0) = \psi\, x(t) = y(t),$$

so that every geodesic is the orbit of a one-parameter group of motions.

Let four points a, a', b, b' with $a\,b = a'\,b' = \delta > 0$ be given and let $x(t), y(t)$ $(0 \leqq t \leqq \delta)$ represent segments $T(a,b)$, $T(a',b')$ with $x(0) = a$, $y(0) = a'$ and extend $x(t)$, $y(t)$ to representations of geodesics. If ψ is a motion taking the first geodesic into the second then

$$\psi\, x(t) = y(\eta\, t + \alpha), \qquad \eta = \pm 1, \text{ for all } t$$

and because ψ is a motion

$$x(t_1)\, x(t_2) = y(\eta\, t_1 + \alpha)\, y(\eta\, t_2 + \alpha) \qquad \text{for any } t_1, t_2.$$

In particular for $t_1 = 0$ and $t_2 = \delta$

$$a\,b = y(\alpha)\, y(\eta\,\delta + \alpha).$$

If Φ_t is a one-parameter group of motions with $y(t) = \Phi_t\, y(0)$ then

$$\Phi_{-\alpha}\, \psi\, a = a', \qquad \Phi_{-\alpha}\, \psi\, b = b' \qquad \text{if } \eta = 1$$

and

$$\Phi_{-\alpha+\delta}\, \psi\, a = b', \qquad \Phi_{-\alpha+\delta}\, \psi\, b = a' \qquad \text{if } \eta = -1,$$

hence

(6) Given four points a, a', b, b' with $a\,a' = b\,b'$ there is a motion Ω such that either $\Omega a = a'$, $\Omega b = b'$ or $\Omega a = b'$, $\Omega b = a'$.

This implies that Γ is transitive on R. Put $K = K(p, \gamma)$ with $0 < \gamma < \rho(p)$. Every point a of K has a unique antipode a' defined by $\gamma = ap = pa' = aa'/2$.

If a, b are given points of K then there is by (6) a motion Ω such that either $\Omega p = p$ and $\Omega a = b$ hence $\Omega \in \Gamma_p$ or $\Omega p = b$ and $\Omega a = p$. In the second case we obtain as before a motion Φ (in a one-parameter subgroup with the geodesic containing $T(p, b)$ as orbit) which takes b into p and p into the antipode b' of b. Then $\Phi \Omega \in \Gamma_p$ and $\Phi \Omega a = b'$.

An orbit of Γ_p as transformation group of K contains therefore at least one of any two antipodes, so that there are not more than two orbits. Two orbits would produce a decomposition of K into two disjoint closed sets, but K is connected, see G (55.1).

Thus Γ_p is transitive on K. Since Γ is transitive on R this means that for any a, a', b, b' with $ab = a'b' = \gamma$ a motion exists which takes a into a' and b into b'. This reduces (5) to (3).

We conclude the discussion of the mobility of geodesics by exhibiting a *remarkable phenomenon:* In the (x, z)-plane consider a function $f(x, z)$ satisfying the conditions:

$$f(x, z) > 0 \quad \text{for } (x, z) \neq (0, 0), \qquad f(kx, kz) = |k|\, f(x, z)$$
$$f(x, z) = f(-x, z), \qquad \text{the curve } C\colon f(x, z) = 1$$

is strictly convex and differentiable, but not an ellipse.

Then the tangents of C at its intersections with the x-axis are parallel to the z-axis.

Let $f^*(x, z) = 1$ $(f^*(kx, kz) = |k|\, f^*(\alpha, z))$ be the polar reciprocal of C revolved through $\pi/2$, compare Section 16.

In the halfspace $z > 0$ of an $(x_1, \ldots, x_{n-1}, z)$-space form the higher-dimensional analogue to quasi-hyperbolic geometry given by the line element

$$ds = z^{-1} f\left(\left(\sum_{i=1}^{n-1} dx_i^2\right)^{\frac{1}{2}}, dz\right).$$

This is invariant under

$$(7) \qquad x_i' = \beta \sum_{i=1}^{n-1} a_{ik} x_k + \alpha_i, \qquad z' = \beta z, \qquad \beta > 0, \ (a_{ik}) \text{ orthogonal}.$$

The space is straight and the geodesics consist of two families F_1, F_2. The halflines $x_i = \text{const}$ $z > 0$ form F_1 and the images of $f^*(x_1, z) = 1, x_2 = \cdots = x_{n-1} = 0, z > 0$ under (7) form F_2.

The mappings (7) form the group Γ of all motions of the space because C is not an ellipse (see, p. 264 of the reference given in footnote 2,

Section 16) *Γ is transitive on the geodesics of either F_i, but will not take a geodesic in F_1 into one of F_2* since (7) consists of linear transformations.

F_2 is dense in the set of all geodesics. This statement presupposes a topology for the geodesics. In general there is no reasonable topology for the geodesics of a G-space (think of transitive geodesics!), however, in a straight space, like the present example, and more generally, when the carriers of the geodesics are closed sets, there is. We may use the distance of the carriers as defined in G, Section 3. In compact G-spaces, in which the geodesics have closed carriers, if the group of motions is transitive on a dense set of geodesics, it is transitive on the set of all geodesics[6]. This shows that the noncompact case offers difficulties which the compact does not.

VI. Observations on Method and Content

Many of the results here and in G may be regarded as contributions to the theory of Finsler spaces. However, few of them are found in the literature based on the standard methods. This work stresses the analogies to the Riemannian case whereas we believe that the attention should be focused on the new phenomena. We will briefly discuss this point.

Our theory deals with problems in the large, but the same difference appears clearly in the simplest local problem. An excellent exposition of the classical local theory up to 1958 can be found in Rund ⟨1⟩. The most interesting feature of this book for us is the detailed analysis of the many connexions proposed for Finsler spaces. They all begin by writing the fundamental metric function $F(x, \xi)$ as

$$(1) \qquad F^2(x,\xi) = \frac{1}{2} \sum_{i,k} \frac{\partial^2 F}{\partial \xi^i \partial \xi^k} \xi^i \xi^k = \sum_{i,k} g_{ik}(x,\xi)\, \xi^i \xi^k$$

and proceed by operating with the g_{ik} which satisfy $g_{ik}(x, \mu\xi) = g_{ik}(x, \xi)$ for $\mu > 0$ and $\xi \neq 0$. Our objections begin at this very first step.

Take the simplest case, a Minkowski plane with a symmetric distance. Compared to the euclidean plane it offers (at least) two essentially new phenomena: 1) the variety of metrics which invites comparison of different ones; 2) through its translations a Minkowski plane determines a natural class of analytic (affine) coordinates, so that a curve of any class is intrinsically defined, whether the metric is degenerate or not. This suggests three postulates concerning the curvature of curves:

(2a) A curve C of class C^∞ (or C^2) possesses at each point p a curvature $\kappa(p)$. For fixed C and p the curvature $\kappa(p)$ varies (2b) continuously and (2c) monotonically with the Minkowski metric[1]. The latter means that $\kappa(p)$ does not decrease if we replace a Minkowski metric by one where all distances are greater. (2b, c), even if restricted to nice metrics, cannot be achieved operating with the g_{ik}. But all three conditions can easily be satisfied if we are willing to use an approach leading to invariants of an entirely different nature, see Busemann [6].

[1] Other examples for comparison of different geometries are contained in the statement: the unit circle of a Minkowski plane is longest (shortest) if it is affinely equivalent to a square (regular hexagon). Notice that $F(x, \xi)$ can in these important cases not be written in the form (1).

It is not our purpose here to develop the local differential geometry of Minkowski spaces, but to point out that valid arguments against basing the theory on the g_{ik} arise as soon as we abandon the preconceived idea that Finsler spaces are to be treated as appendages to Riemann spaces.

Returning briefly to connexions:

The first systematic theory evolved in the then Prague school of Funk, Koschmieder, Winternitz and, principally, Berwald. This direction was continued by Varga until his recent death (1969). The impetus came from Hilbert's Problem IV to investigate the Desarguesian spaces and the theory contains many attractive contributions in this area, several of which were mentioned here (Sections 2 and 8).

Then Cartan ⟨1⟩ proposed a strictly Riemannian approach and obtained through certain postulates a definite euclidean connexion which, in contrast to Berwald's, satisfies Ricci's Lemma. For a while it was thought that this connexion would supersede all others, but, as emphasized by Rund, it did not because it is poorly adapted to questions of a non-Riemannian character. Chern ⟨1⟩ developed a general theory of euclidean connexions in Finsler spaces which is not incorporated in Rund ⟨1⟩.

All connexions which were discussed hitherto as well as others not mentioned here (see Rund ⟨1, Chapter II⟩) are linear. There is also a general theory of nonlinear connexions by Vagner, for references see his comprehensive paper ⟨1⟩ and Rund ⟨1⟩. Vagner studied fields of local hypersurfaces in A^n. In the applications to Finsler spaces these are the local unit spheres $F(x, \xi)=1$. An important special case is the connexion of Barthel ⟨1⟩ leading to a parallel displacement which preserves the Minkowski lengths of vectors.

The multiplicity of connexions and the fact that our approach does not use any while containing by far the largest body of theorems on Finsler spaces in the large, raises doubts concerning the actual importance of connexions in this area.

It would not be fair to make this statement without emphasizing that some important Riemannian theorems have been extended to Finsler spaces by Auslander ⟨1⟩ using Chern's theory and more recently by Moalla ⟨1⟩ based on Cartan.

Both authors supplement the original methods by modern concepts like fibre spaces and operate in the tangent bundle.

With not quite identical concepts Bonnet's Theorem (see Section 17) is extended to n dimensions in either paper under the assumption that an analogue to the Ricci curvature exceeds a positive constant[2]. Moalla

[2] In the Riemannian case the theorem in this form is due to Myers ⟨1⟩.

also shows that the fundamental group is finite. Auslander generalizes Synge's Theorem that a compact even-dimensional space with positive curvature is simply connected. The fact that a simply connected space with nonpositive curvature is homeomorphic to E^n (corresponding to the fact that a simply connected space with convex capsules and domain invariance is straight) is found in either paper. Moalla shows that spaces with nonpositive sectional curvature in his sense have *convex* capsules. He also defines a scalar curvature and proves that a compact space without conjugate points has nonpositive scalar curvature.

Certain topics like the existence of transitive geodesics on certain surfaces have so natural a formulation in terms of the tangent bundle that one would be willing to grant that they must readily yield to these methods, although neither author treats this question. Other problems involving the topology of geodesics as pointsets in the given space, or geodesic polygons, seem less accessible, and one does not see at all how these methods would apply to typically non-Riemannian situations, for example our theorems (7.1) and (15.10, 14).

The mentioned papers and others confirm that much of Riemannian geometry, and in particular most of the theorems in the large with an intuitive geometric content, do not depend on the Riemannian character of the metric. Therefore, coming back to our original point, we think that emulation of Riemannian methods or results is not the principal concern of Finsler geometry. A contribution to Finsler spaces which does not deal with a problem nonexistent or different in the Riemannian case should either be new also in this case or at least contain a substantial generalization of a known fact (like the generalization from 2 to n dimensions or from compact to complete surfaces in our theorems (12.2) resp. (13.6)). Typically non-Riemannian situations are offered by Hilbert's Problem IV (Theorem (7.1) is a contribution) and by the plane inverse problems, to which we devoted so much attention just because they fall into this category. Last but not least, we should probe into the limits of Riemannian geometry. The ideal, quite utopian at present, would be a criterion which allows us to determine which statements are truly Riemannian. Failing this we must try to gain understanding step by step. The nonexistence of a Bonnet angle (Section 16) is an isolated example and it is not clear whether our proof contains ideas applicable to other situations. The deep and penetrating questions lie, in our opinion, in this direction, because some basic insight must be missing in Riemannian geometry if we are never certain whether one of the principal hypotheses, the quadratic character of the line element, is relevant.

Not only the content but also the methods in our theory differ from the usual ones because we omit differentiability hypotheses. In some cases this is, of course, necessary as in the investigation of differentiability itself or of equilong maps in Section 11, where all interesting maps are nondifferentiable. In others, like spaces with convex capsules, one would not like to miss the insight that nonpositive curvature is expressible without differentiability.

But we do not claim that the irrelevance of differentiability in much of differential geometry, although it sounds very startling, leads in our context to problems which are in any sense as deep as those mentioned above[3]. Our approach establishes a link with the foundations of geometry which we find very appealing and which is quite conspicuous in Hilbert's thinking, but has all but disappeared in contemporary mathematics. Our axioms extract the essential geometric properties from Finsler spaces (with a symmetric distance in the case of G-spaces). Seen in this light their effectiveness is not surprising and nothing basic appears to be lacking[4].

We do like intuitive geometric arguments and uncovering simple geometric reasons underlying seemingly recondite facts. But this is a matter of taste.

In the last few years spaces with indefinite metrics have attracted increasing attention, so that the reader may want to know whether the present methods have been applied to this area. Such questions were taken up lately by the author in ⟨11⟩ and in Busemann and Beem ⟨1⟩. A brief report can be found in our lecture ⟨12⟩. However, the latter does not include the recent work of Beem and Woo ⟨1⟩, who obtained far reaching results for two dimensions.

Note

After completion of the manuscript J. D. Featherstone[5] and B. B. Phadke[6] began a systematic investigation of the spaces with a not (necessarily) symmetric intrinsic distance, the prolongation properties P and U of Section 2 and the weak completeness condition that $\bar{S}^{+}(p,\rho)$ be compact.

As predicted in the preface major changes proved necessary. Two of the easily stated elementary phenomena causing frequent difficulties are these:

---

[3] It does in others, like the work of A.D. Alexandrov and Pogorelov.

[4] In the contrary, there are theorems with far from trivial implications for Riemann spaces in which our axioms are or would be unnecessarily strong. (12.2) closed hyperbolic space forms is an example and most probably not the only one.

[5] Spaces with nonsymmetric distance and compactness. Dissertation University of Southern California 1970. The principal results will be published.

[6] Equidistant loci. Dissertation University of Southern California 1970. The results will be published.

1) If $x_\nu(t)$ ($\nu=0,1,2,\ldots$; $\alpha_\nu<t<\infty$) represent maximal partial geodesics and $x_\nu(t)\to x_0(t)$ for $|t|\leqq\delta$ ($\delta>0$), then in general not $\alpha_\nu\to\alpha_0$, only $\limsup\alpha_\nu\geqq\alpha_0$.

2) If M is a closed set, then a given point p possesses a "terminal foot" $f\in M$ with $pf=\min_{x\in M} px$, but not always an "initial foot" $f'\in M$ with $f'p=\min_{x\in M} xp$.

The results of Sections 4, 5 as well as other basic properties can be extended, although with some nontrivial complications. Featherstone developed in particular the theory of parallels and of spaces with convex capsules, which requires complete revamping. Phadke discussed equidistant loci. A noncompact desarguesian space (see Theorem (8.2)) is Minkowskian if for each hyperplane H and $\alpha>0$ and the halfspaces H_1, H_2 bounded by H the loci $E_i^+=\{x|x\in H_i,\, xH=\alpha\}$ and $E_i^-=\{x|x\in H_i,\, Hx=\alpha\}$ ($i=1,2$) are hyperplanes or, if for each line L and $\alpha>0$ the sets $C^+(L,\alpha)=\{x|xL\leqq\alpha\}$ and $C^-(L,\alpha)=\{x|Lx\leqq\alpha\}$ are convex and unions of lines. (It does not suffice to assume, for example, that for each H and α one of the $E_i^\pm(H,\alpha)$ be a hyperplane.) The proofs are very long but would materially reduce under the – among contributors to Hilbert's Problem IV common – assumption that the space be either all of A^n or bounded by a strictly convex hypersurface.

Literature[1]

Alexandrov, A. D.
⟨1⟩ Über eine Verallgemeinerung der Riemannschen Geometrie. Ber. Riemann Tagung Forschungsinst. für Math. Heft 1, 33–84 (1957).

Alexandrov, A. D., Zalgaller, V. A.
⟨1⟩ Two-dimensional manifolds with bounded curvature [Russian]. Trudy Math. Inst. Steklov No. 63 (1962); English translation in Proc. Steklov Inst. Math. No. 76 (1966).

Ambrose, W., Singer, J. M.
⟨1⟩ A theorem on holonomy. Trans. Amer. Math. Soc. **75**, 428–443 (1953).[2]

Auslander, L.
⟨1⟩ On curvature in Finsler geometry. Trans. Amer. Math. Soc. **79**, 378–388 (1955).

Barthel, W.
⟨1⟩ Über eine Parallelverschiebung mit Längeninvarianz in lokal-Minkowskischen Räumen I, II. Arch. Math. **4**, 346–365 (1953).

Beem, J., Woo, P.
⟨1⟩ Doubly timelike surfaces. Mem. Amer. Math. Soc., No. 92 (1969).

Berwald, L.
⟨1⟩ Über eine charakteristische Eigenschaft der allgemeinen Räume konstanter Krümmung mit gradlinigen Extremalen. Monatsh. Math. Phys. **36**, 315–330 (1929).
⟨2⟩ Parallelübertragung in allgemeinen Räumen. Atti Congr. Mat. Bologna IV, 263–270 (1928).
⟨3⟩ Über Finlersche und Cartansche Geometrie I. Mathematica **12**, 34–58 (1941).[3]

Bishop, R. L., Crittenden, R. J.
⟨1⟩ Geometry of manifolds. New York 1964.

Blaschke, W., Bol, G.
⟨1⟩ Geometrie der Gewebe. Berlin 1938.

Bliss, G. A.
⟨1⟩ The geodesics on an anchor ring. Ann. of Math. **4**, 1–20 (1903).

Bochner, S.
⟨1⟩ Vector fields and Ricci curvature. Bull. Amer. Math. Soc. **52**, 776–797 (1946).

---

[1] This list contains only items not found in Busemann ⟨1⟩, see preface.

[2] This is the correct reference for footnote 1, *G* Section 49, see *G* p. 415.

[3] Quoted in *G* p. 271, as Berwald [1], but omitted in the Literature *G* p. 409.

Borsuk, K.
⟨1⟩ Quelques théorèmes sur les ensembles unicohérents. Fund. Math. **17**, 171–209 (1931).
⟨2⟩ Über sphäroidale and H-sphäroidale Räume. Recueil Math. **1**, 643–650 (1936).

Busemann, H.
⟨1⟩ The geometry of geodesics. New York 1955, quoted as G.
⟨2⟩ On normal coordinates in Finsler spaces. Math. Ann. **129**, 417–423 (1955).
⟨3⟩ Groups of motions transitive on sets of geodesics. Duke Math. J. **24**, 539–544 (1956).
⟨4⟩ Metrizations of projective spaces. Proc. Amer. Math. Soc. **8**, 387–390 (1957).
⟨5⟩ Similarities and differentiability. Tôhoku Math. J. (II) **9**, 56–67 (1957).
⟨6⟩ Spaces with finite groups of motions. J. Math. Pures Appl. (9) **37**, 365–373 (1958).
⟨7⟩ On spaces with one-parameter groups of motions. Tensor (New Ser.) **10**, 21–25 (1960).
⟨8⟩ Geometries in which the planes minimize area. Ann. Mat. Pura Appl. (IV) **55**, 171–190 (1961).
⟨9⟩ Length-preserving maps. Pacific J. Math. **14**, 457–477 (1964).
⟨10⟩ Extremals on closed hyperbolic space forms. Tensor (New Ser.) **16**, 313–318 (1965).
⟨11⟩ Timelike spaces. Dissert. Math. No. 53, Warsaw, 1967.
⟨12⟩ Synthetische Differentialgeometrie. Jber. Deutsch. Math.-Verein. **71**, 1–24 (1968).

Busemann, H., Beem, J. K.
⟨1⟩ Axioms for indefinite metrics. Rend. Circ. Mat. Palermo (Ser. II) **15**, 223–246 (1966).

Busemann, H., Pedersen, F. P.
⟨1⟩ Tori with one-parameter groups of motions. Math. Scand. **3**, 209–220 (1955).

Busemann, H., Salzmann, H.
⟨1⟩ Metric collineations and inverse problems. Math. Z. **87**, 214–240 (1965).

Carathéodory, C.
⟨1⟩ Variationsrechnung und partielle Differentialgleichungen. Leipzig u. Berlin 1935.
⟨2⟩ Gesammelte mathematische Schriften, vol. I. München 1964.

Cartan, E.
⟨1⟩ Les Espaces de Finsler. Actualités Sci. Indust. (Paris) **79** (1934).

Chern, S. S.
⟨1⟩ On the euclidean connections in a Finsler space. Proc. Nat. Acad. Sci. U.S.A. **29**, 33–37 (1943).

Cohn-Vossen, S.
⟨1⟩ Existenz kürzester Wege. Dokl. Akad. Nauk SSSR **3**, 339–342 (1935).

Darboux, G.
⟨1⟩ Leçons sur la théorie générale des surfaces, vol. III. Paris 1894.

Dugundji, J.
⟨1⟩ Topology. Boston 1966.

Efremovič, V.A., Tihomirova, E.S.
⟨1⟩ Equimorphisms of hyperbolic spaces [Russian]. Izv. Akad. Nauk SSSR, Ser. Mat. **28**, 1139–1144 (1964).

Eilenberg, S., Steenrod, N.
⟨1⟩ Foundations of algebraic topology. Princeton 1952.

Freudenthal, H.
⟨1⟩ Neuere Fassungen des Riemann-Helmholtz-Lieschen Raumproblems. Math. Z. **63**, 374–405 (1956).
⟨2⟩ Lie groups in the foundations of geometry. Advances in Math. **1**, 145–190 (1965).

Freudenthal, H., Hurewicz, W.
⟨1⟩ Dehnungen, Verkürzungen, Isometrien. Fund. Math. **26**, 120–122 (1936).

Funk, P.
⟨1⟩ Über Geometrien, bei denen die Geraden die Kürzesten sind. Math. Ann. **101**, 226–237 (1929).
⟨2⟩ Über zweidimensionale Finslersche Räume, insbesondere solche mit gradlinigen Extremalen und positiver konstanter Krümmung. Math. Z. **40**, 86–93 (1935).

Green, L.W.
⟨1⟩ Auf Wiedersehensflächen. Ann. of Math. **78**, 289–299 (1963).

Hausdorff, F.
⟨1⟩ Dimension und äußeres Maß. Math. Ann. **79**, 157–179 (1919).

Hewitt, E., Stromberg, K.
⟨1⟩ Real and abstract analysis. New York 1965.

Hilbert, D.
⟨1⟩ Grundlagen der Geometrie, 8th edit. Stuttgart 1956.

Hopf, H., Rinow, W.
⟨1⟩ Über den Begriff der vollständigen differentialgeometrischen Fläche. Comment. Math. Helv. **3**, 209–225 (1932).

Kann, E.
⟨1⟩ Bonnet's theorem in two-dimensional *G*-spaces. Comm. Pure Appl. Math. **14**, 765–784 (1961).

Kelly, P., Straus, E.G.
⟨1⟩ Curvature in Hilbert geometry. Pacific J. Math. **8**, 119–125 (1958).

Kimball, B.F.
⟨1⟩ Geodesics on a toroid. Amer. J. Math. **52**, 29–52 (1930).

Kirk, W.A.
⟨1⟩ On locally isometric mappings of a *G*-space on itself. Proc. Amer. Math. Soc. **15**, 584–586 (1964).
⟨2⟩ Isometries in *G*-spaces. Duke Math. J. **31**, 539–541 (1964).
⟨3⟩ A theorem on local isometries. Proc. Amer. Math. Soc. **17**, 453–455 (1966).

Kosiński, A.
⟨1⟩ On manifolds and r-spaces. Fund. Math. **42**, 119–125 (1955).

Krakus, B.
⟨1⟩ Any 3-dimensional G-space is a manifold. Bull. Acad. Polon. Sci. Sér. Math. Astronom. Phys. **16**, 737–740 (1968).

Lichnerowicz, A.
⟨1⟩ Sur une extension de la formule d'Allendoerfer-Weil à certaines variétés finslériennes. C. R. Acad. Sci. Paris **223**, 12–14 (1946).

Löbell, F.
⟨1⟩ Beispiele geschlossener dreidimensionaler Clifford-Kleinscher Räume negativer Krümmung. Ber. Verh. Sächs. Akad. Wiss. Leipzig Math.-Phys. Kl. **83**, 167–174 (1931).

Moalla, F.
⟨1⟩ Sur quelques théorèmes globaux en géométrie finslérienne. Ann. Mat. Pura Appl. (4) **73**, 319–365 (1966).

Montgomery, D., Zippin, L.
⟨1⟩ A theorem on the rotation group of the two-sphere. Bull. Amer. Math. Soc. **46**, 520–521 (1940).

Myers, S. B.
⟨1⟩ Riemannian manifolds in the large. Duke Math. J. **31**, 33–49 (1935).

Nasu, J.
⟨1⟩ On asymptotic conjugate points. Tôhoku Math. J. Ser. 2, **7**, 157–165 (1955).
⟨2⟩ On asymptotes in metric spaces with non-positive curvature. Tôhoku Math. J. Ser. 2, **9**, 68–95 (1957).
⟨3⟩ On asymptotes in two-dimensional metric spaces. Tensor New Ser. **7**, 173–184 (1957).

Rinow, W.
⟨1⟩ Die innere Geometrie der metrischen Räume. Berlin-Göttingen-Heidelberg 1961.

Rolfsen, D.
⟨1⟩ Strongly convex metrics in cells. Bull. Amer. Soc. **74**, 171–175 (1968).

Rund, H.
⟨1⟩ The differential geometry of Finsler spaces. Berlin-Göttingen-Heidelberg 1959.

Salzmann, H. R.
⟨1⟩ Topological planes. Advances in Math. **2**, 1–60 (1967).

Schoenberg, I. J.
⟨1⟩ Some applications of the calculus of variations to Riemannian geometry. Ann. of Math. **33**, 485–495 (1932).

Skornyakov, L. A.
⟨1⟩ Metrization of the projective plane in connection with a given system of curves [Russian]. Ivz. Akad. Nauk SSSR Ser. Mat. **19**, 471–482 (1955).

Spanier, E.H.
⟨1⟩ Algebraic topology. New York 1966.

Szenthe, J.
⟨1⟩ Über ein Problem von H. Busemann. Publ. Math. Debrecen **7**, 408–413 (1960).
⟨2⟩ Über lokalisometrische Abbildungen von *G*-Raumen auf sich. Ann. Mat. Pura Appl. (IV **55**, 37–46 (1961).
⟨3⟩ Über metrische Räume, deren lokalisometrische Abbildungen Isometrien sind. Acta Math. Acad. Sci. Hungar. **13**, 433–441 (1962).

Tits, J.
⟨1⟩ Sur un article précédent: Études de certains espaces métriques. Bul. Soc. Math. Belg. 1953 (1953), 124–128.
⟨2⟩ Sur certaines classes d'espaces homogènes de groupes de Lie. Mem. Acad. Roy. Belg. Sci. **29**, fasc. 3 (1955).

Vagner, V.V.
⟨1⟩ The geometry of Finsler spaces as a theory of a field of local hypersurfaces in X_n [Russian]. Trudy Sem. Vektor. Tenzor. Anal. **7**, 65–166 (1949).

Wolf, J.A.
⟨1⟩ Spaces with constant curvature. New York 1966.

Zaustinsky, E.M.
⟨1⟩ Spaces with non-symmetric distance. Mem. Amer. Math. Soc. No. 34 (1959).
⟨2⟩ Extremals on compact *E*-surfaces. Trans Amer. Soc. **102**, 433–445 (1962).

Ergebnisse der Mathematik und ihrer Grenzgebiete

1. Bachmann: Transfinite Zahlen. DM 38,–; US $ 10.50
2. Miranda: Partial Differential Equations of Elliptic Type. DM 58,–; US $ 16.00
4. Samuel: Méthodes d'Algèbre Abstraite en Géométrie Algébrique. DM 26,–; US $ 7.20
5. Dieudonné: La Géométrie des Groupes Classiques. DM 38,–; US $ 10.50
6. Roth: Algebraic Threefolds with Special Regard to Problems of Rationality. DM 19,80; US $ 5.50
7. Ostmann: Additive Zahlentheorie. 1. Teil: Allgemeine Untersuchungen. DM 38,–; US $ 10.50
8. Wittich: Neuere Untersuchungen über eindeutige analytische Funktionen. DM 28,–; US $ 7.70
11. Ostmann: Additive Zahlentheorie. 2. Teil: Spezielle Zahlenmengen. DM 28,–; US $ 7.70
14. Coxeter/Moser: Generators and Relations for Discrete Groups. DM 32,–; US $ 8.80
16. Cesari: Asymptotic Behavior and Stability Problems in Ordinary Differential Equations. DM 36,–; US $ 9.90
17. Severi: Il teorema di Riemann-Roch per curve-superficie e varietà questioni collegate. DM 23,60; US $ 6.50
18. Jenkins: Univalent Functions and Conformal Mapping. DM 34,–; US $ 9.40
19. Boas/Buck: Polynomial Expansions of Analytic Functions. DM 16,–; US $ 4.40
20. Bruck: A Survey of Binary Systems. DM 36,–; US $ 9.90
21. Day: Normed Linear Spaces. DM 17,80; US $ 4.90
25. Sikorski: Boolean Algebras. DM 38,–; US $ 9.50
26. Künzi: Quasikonforme Abbildungen. DM 39,–; US $ 10.80
27. Schatten: Norm Ideals of Completely Continuous Operators. DM 26,–; US $ 7.20
28. Noshiro: Cluster Sets. DM 36,–; US $ 9.90
29. Jacobs: Neuere Methoden und Ergebnisse der Ergodentheorie. DM 49,80; US $ 13.70
30. Beckenbach/Bellman: Inequalities. DM 30,–; US $ 8.30
31. Wolfowitz: Coding Theorems of Information Theory. DM 27,–; US $ 7.50
32. Constantinescu/Cornea: Ideale Ränder Riemannscher Flächen. DM 68,–; US $ 18.70
33. Conner/Floyd: Differentiable Periodic Maps. DM 26,–; US $ 7.20
34. Mumford: Geometric Invariant Theory. DM 22,–; US $ 6.10
35. Gabriel/Zisman: Calculus of Fractions and Homotopy Theory. DM 38,–; US $ 9.50
36. Putnam: Commutation Properties of Hilbert Space Operators and Related Topics. DM 28,–; US $ 7.70
37. Neumann: Varieties of Groups. DM 46,–; US $ 12.60
38. Boas: Integrability Theorems for Trigonometric Transforms. DM 18,–; US $ 5.00
39. Sz.-Nagy: Spektraldarstellung linearer Transformationen des Hilbertschen Raumes. DM 18,–; US $ 5.00
40. Seligman: Modular Lie Algebras. DM 39,–; US $ 9.75
41. Deuring: Algebren. DM 24,–; US $ 6.60
42. Schütte: Vollständige Systeme modaler und intuitionistischer Logik. DM 24,–; US $ 6.60
43. Smullyan: First-Order Logic. DM 36,–; US $ 9.90
44. Dembowski: Finite Geometries. DM 68,–; US $ 17.00
45. Linnik: Ergodic Properties of Algebraic Fields. DM 44,–; US $ 12.10
46. Krull: Idealtheorie. DM 28,–; US $ 7.70
47. Nachbin: Topology on Spaces of Holomorphic Mappings. DM 18,–; US $ 5.00

48. A. Ionescu Tulcea/C. Ionescu Tulcea: Topics in the Theory of Lifting. DM 36,– ; US $ 9.90
49. Hayes/Pauc: Derivation and Martingales. DM 48.– ; US $ 13.20
50. Kahane: Séries de Fourier Absolument Convergentes. DM 44,– ; US $ 12.10
51. Behnke/Thullen: Theorie der Funktionen mehrerer komplexer Veränderlichen. DM 48,– ; US $ 13.20
52. Wilf: Finite Sections of Some Classical Inequalities. DM 28,– ; US $ 7.70
53. Ramis: Sous-ensembles analytiques d'une variété banachique complexe. DM 36,– ; US $ 9.90
54. Busemann: Recent Synthetic Differential Geometry. DM 32,– ; US $ 8.80
55. Walter: Differential and Integral Inequalities. In preparation